Advanced Analytics in Power BI with R and Python

Ingesting, Transforming, Visualizing

Ryan Wade

Apress®

Advanced Analytics in Power BI with R and Python: Ingesting, Transforming, Visualizing

Ryan Wade
Indianapolis, IN, USA

ISBN-13 (pbk): 978-1-4842-5828-6   ISBN-13 (electronic): 978-1-4842-5829-3
https://doi.org/10.1007/978-1-4842-5829-3

Managing Director, Apress Media LLC: Welmoed Spahr
Acquisitions Editor: Jonathan Gennick
Development Editor: Laura Berendson
Coordinating Editor: Jill Balzano

Cover image designed by Freepik (www.freepik.com)

Distributed to the book trade worldwide by Springer Science+Business Media New York, 233 Spring Street, 6th Floor, New York, NY 10013. Phone 1-800-SPRINGER, fax (201) 348-4505, e-mail orders-ny@springer-sbm. com, or visit www.springeronline.com. Apress Media, LLC is a California LLC and the sole member (owner) is Springer Science + Business Media Finance Inc (SSBM Finance Inc). SSBM Finance Inc is a **Delaware** corporation.

For information on translations, please e-mail booktranslations@springernature.com; for reprint, paperback, or audio rights, please e-mail bookpermissions@springernature.com.

Apress titles may be purchased in bulk for academic, corporate, or promotional use. eBook versions and licenses are also available for most titles. For more information, reference our Print and eBook Bulk Sales web page at http://www.apress.com/bulk-sales.

Any source code or other supplementary material referenced by the author in this book is available to readers on GitHub via the book's product page, located at www.apress.com/9781484258286. For more detailed information, please visit http://www.apress.com/source-code.

Printed on acid-free paper

I'd like to thank my colleagues at BlueGranite. It is an honor to be part of a team of very smart and talented individuals. Iron sharpens Iron. I'd like to thank the neighborhood I grew up in, the west side of Michigan City, IN. My experience growing up there is priceless. I want to thank Patrick Leblanc, Mico Yuk, Terry Morris, Dr. Brandeis Marshall, and Dr. Sydeaka Watson. Each of you has attributes that I admire much, and you all mentor me from afar via your examples.

I'd like to thank all my teammates from little league to college. Athletics has been one of my best life teachers, and it was an honor to go through the process with you all. I look up to and respect many of you. I'd like to thank my former coaches, especially my high-school and collegiate coaches. The tough challenges you placed me in caused me to develop grit that has helped me in other aspects of my life. Because of that, I will always be indebted to you.

I'd like to thank my extended family. It takes a village to raise a child, and I have benefited from the support of many family members, and I appreciate that. I'd like to thank my mentees, Tim Adams Jr. and Camille Little. Both of you have a rare combination of very high intellect and very likable personalities. In a short time, roles will switch, and you will become the mentor, and I will be the mentee! I'd like to thank my brothers from Michigan City and my brothers that I met while playing football in Louisville. I experienced blood, sweat, and tears with you all. We jumped off the porch and became men together. No matter what, we will be brothers for life! I'd like to thank my first cousins. We grew up like brothers and sisters and shared many fond memories. Some of you had my back in a time of need. That will never be forgotten. I want to thank my siblings, Paulette (RIP), Stephanie, Tina, and Luke, and my nephew Junebug, who was raised like my brother. When we needed each other, we always had each other's back. Let's stay that way. I'd like to thank my mom and dad, Luther and Ernestine Wade. I appreciate your love, support, and sacrifice for me and my siblings. It is much appreciated! In your words, daddy, "I love you, and that is always."

Last but not least, I'd like to thank God. I appreciate all the talents you have given me. I will show my appreciation by using them to the fullest so that I can be a benefit to my community.

Table of Contents

Part II: Ingesting Data into the Power BI Data Model Using R and Python

About the Author

Ryan Wade is a data analytic professional with over 20 years of experience. His education and work experience enable him to have a holistic view of analytics from a technical and business viewpoint. He has an MCSE with an emphasis on BI reporting and Microsoft R. He has an advanced understanding of R, Python, DAX, T-SQL, M, and VBA. He knows how to leverage those programming languages for on-prem and cloud-based data analytics solutions using the Microsoft Data Platform.

Ryan is a data analytics enthusiast, and he has spoken at R meetups, Python meetups, SQLSaturdays, TDWI Conference, BDPA Conference, and PASS Summit about various data analytics topics. He is the developer of a comprehensive online course for ExcelTv showing how to implement R in Power BI for advanced data analytics and data visualization.

About the Technical Reviewer

Aaditya Maruthi works as a Senior Database Engineer for a reputed organization. Having over ten years of experience in RDMS systems like Microsoft SQL Server and Oracle, he worked extensively on Microsoft technologies like the SSAS, SSRS, SSIS, and Power BI.

Aaditya is also a Certified AWS Solutions Architect Associate.

Acknowledgments

I want to thank the technical editors, Mike Huffer and Aaditya Maruthi. The feedback that you all gave me was very valuable and much appreciated. I'd also like to thank Jonathan Gennick and Jill Balzano. I appreciate your patience and help. I would not have been able to complete the process without your guidance. I also want to thank both of you for keeping things in perspective. There was so much I wanted to include, but that would have taken way too long to write. You helped me decide what was important, which helped us finish the book in a reasonable time.

Introduction

Microsoft Power BI is considered by many to be the premier self-service business intelligence tool on the market. In recent years, it has passed up formidable tools such as QlikView and Tableau to gain the number one spot. One of the reasons why it is considered such a great self-service business intelligence tool is because it is more than just a visualization tool. Built into Microsoft Power BI are

- The *DAX* expression and query language that enables you to interrogate your Power BI data model using complex business logic in a fast and efficient way

- A data wrangling tool called *Power Query* that enables you to shape and transform your data into a form that is conducive for data analysis

- The *Vertipaq* engine that efficiently stores the data in a way that is optimized for reporting and performs complex calculations fast and efficiently

- Pre-packaged, interactive visualizations that enable you to present your data in ways that are easily understood by report consumers

So, you may ask the question, with all these features why would you need to leverage programming languages such as R and Python in Power BI? The answer is to fill in the few areas where the native tools fall short. A few examples are

- Creating custom visualizations in a relatively easy way

- Applying data science to your Power BI data models without the need of *Power BI Premium*

- Performing advanced string manipulations using advanced techniques that are not available in *Power Query* or *DAX*

- Interacting with *Microsoft Cognitive Services* without the need of *Power BI Premium*

- Communicating with third-party data APIs to enrich your Power BI data models in an efficient way

- And many more

This book covers how to leverage R and Python to bring the added functionality listed earlier to your Power BI solutions. R is a perfect complement to Power BI because it is a language written specifically for data analytics. Data analysts have been using R to perform tasks like data wrangling and data visualization for decades. Given that, features that may not be available in Power BI might have been in R for some time.

Python has become a very popular programming language in data analytics over the last decade. One of the features that make Python so attractive is that it is not only great for data analytics, but it is great for general programming tasks as well. Communicating with APIs is a breeze with Python, but the same task is very clunky using Power Query.

These features of R and Python make them perfect companions to Power BI. This book will cover some recipes that illustrate the preceding features. The recipes will include detailed steps along with verbose descriptions so that you will get a clear understanding of how they work. Before you get started using the recipes, you need to configure your environment. Let's go over those configurations.

Configure your Azure environment

Different parts of *Azure* will be used throughout the book. *Microsoft Cognitive Services* will be used to apply artificial intelligence to your Power BI data models. Also, the *Data Science Virtual Machine (DSVM)* is the recommended development environment for the book. The *DSVM* is optional but highly recommended. In order to work many of the examples in this book, you will need an environment configured with many tools such as

- SQL Server 2017 or later

- SQL Server Machine Learning Services 2017 or later with R and Python enabled

- The Anaconda distribution of Python

- The R programming language

- R Studio

- VS Code

- Power BI Desktop

Manually configuring an environment with these resources can be a challenge, but if you use the *DSVM*, most of the configuration is handled for you. In the following sections, you will learn how to set up *Azure* so that you can consume *Microsoft Cognitive Services* and you will also learn how to spin up a DSVM.

Sign up for Azure

Sign up for Azure here: `https://azure.microsoft.com/en-us/free/`. You get 12 months free for selected services plus $200 in credit during your first month!

Sign up for Microsoft Cognitive Services

Microsoft Cognitive Services will be used to perform sentiment analysis inside of Power BI using Python. You first need to set up the *Microsoft Cognitive Services* in *Azure* before you can call it from Power BI. Here are the required steps to set up the service:

1. Log in to your Azure Portal.

2. Type *Cognitive Services* in the search box, then press Enter.

3. The preceding action should take you to the *Cognitive Service* signup page. Click the *Create* button to initiate the signup process.

4. Fill in the following information:

   - Name

   - Subscription

   - Location

   - Pricing tier

   - Resource group

5. Click the *I confirm I have read and understood the notice below* check box.

6. Click the *Create* button.

Note that there is a cost to use *Microsoft Cognitive Services*. To get pricing information, go to your *Microsoft Cognitive Services* resource and type *Pricing tier* in the search box, then click it in the results. Select the pricing tier you want to use. Doing so will take you to a page that will give you pricing information based on your usage and *Azure* region. The exercise in this book that uses *Microsoft Cognitive Services* is relatively inexpensive, and you will have more than enough credits in your first month to cover the cost.

Create a Data Science Virtual Machine (DSVM)

The preferred setup is to start with a *Data Science Virtual Machine (DSVM)* and add the resources that do not come pre-installed in the *DSVM* to it. This setup is highly preferred because the amount of time it takes to fully configure your environment in the way that is needed to do every exercise in the book can be lengthy and challenging. Using the *DSVM* enables you to configure an environment in minutes that could take you many days of trial and error if you tried to do it yourself. Instructions will be provided in the next section if you decide to configure your own environment. Here are the instructions to set up the *DSVM*.

Steps to create a DSVM in Azure

1. Go to `https://portal.azure.com`. If prompted, sign in using the credentials created in Step 1.

2. Click *Create a resource* in the upper left.

3. In the search box, type *Data Science Virtual Machine - Windows 2019.*

4. Click the *Create* button. This action will cause a form to appear that you need to fill out to configure the DSVM. You will land on the *Basics* tab. The following steps will tell you how to fill out the form.

5. *Subscription*: Select the subscription you want to use. It should default to the subscription that you set up in Step 1.

6. *Resource group*: If you already have a resource group in your tenant you want to use, select it. Otherwise, create a new one for your DSVM">.

7. *Virtual machine name*: The name you want your DSVM to have.

8. *Region*: An Azure region close to you.

9. *Image*: Make sure Data Science Virtual Machine – Windows 2016 is selected.

10. *Size*: I use B4ms because it is the cheaper option for the 16 gigs of ram options. RAM is important for R, Python, and Power BI.

11. *Username*: <"Create a username">.

12. *Password*: <"Create password">.

13. *Confirm password*: <"Confirm password">.

14. Click *Next: Disks >*.

15. *OS disk type*: Select Standard SSD. This option is sufficient for what we are doing.

16. Click *Next: Networking>*.

17. Make sure all the required fields are filled out. You can identify the required fields with a *. They should be populated with a default value; if not, click the Create New link below them and accept the default settings.

18. Click *Next: Management >*.

19. Accept defaults and click *Next: Advanced >*.

20. Accept defaults and click *Next: Tags >*.

21. Click *Next: Review + create >*.

22. You will see a summary of the DSVM that you configured. You will also see the cost to run the DSVM. If you agree with the configuration, click the *Create* button.

Make sure to stop your DSVM after every use so that you don't incur unnecessary costs. As a safeguard, you should set up an auto-shutdown that will shut down your *DSVM* if it is still running past a specified time. Do the following steps to set up auto-shutdown:

1. Go into your *DSVM* machine in the *Azure Portal*.

2. Type *auto-shutdown* in the search box, then select it in the list.

3. Enable auto-shutdown by selecting *On* in the *Enabled* button.

4. Select the time you would like to automate the shutdown in the *Scheduled shutdown* textbox.

5. Choose the time zone you want to base the time on in the *Time zone* combo box.

6. If you want notification to be sent to you that lets you know that your DSVM will be shutting down, you can turn on the *Send notification before auto-shutdown.* The notification will be sent to the email that you put in the *Email address* textbox.

Configure R in DSVM

You will be using a different distribution of R than the one installed. The distribution of R that we will use is *Microsoft R Open (MRO).* This distribution of R is totally compatible with distribution on CRAN, but it comes with enhancements that improve the performance of certain types of calculations plus many additional tools. Perform the following steps to download the *MRO* in your *DSVM*:

1. Get the version of R that is being used in the Power BI service. You can find that information in the following Microsoft documentation: `https://docs.microsoft.com/en-us/power-bi/visuals/service-r-visuals`.

2. Open up a browser in the *DSVM* and go to the following site: `https://mran.microsoft.com/open`. Two browsers are pre-installed in the DSVM. They are *Microsoft Edge* and *Firefox*.

3. Click the *Download* button on the right and you will be taken to the download page. Once on the download page, click the *Past Releases* link which is located on the right section of the page. Clicking the link will take you to a page that has links to all previous versions of *Microsoft R Open*. Click the link for the version that the Power BI service is using.

4. Choose the download for *Windows*.

5. Execute the download.

6. Open R Studio in the *DSVM*.

7. Select *Tools* ➤ *Global Options*. Verify that the MRO distribution you just installed is selected. If it is not, click the *Change...* button and you should see it as one of the options. Select it, then click *OK*.

Configure Python in DSVM

One of the benefits you gain with the DSVM is you get a distribution of Python pre-installed that is perfect for data analytics. The name of the distribution is *Anaconda*. The *Anaconda* distribution of Python comes pre-installed with over 1500 libraries that are popular in data analytics. It also comes with a package manager and environment management system named *conda*. Installing libraries via *conda* is preferred because of how *conda* manage package dependencies. The environment management system in *conda* makes it easy to create an isolated copy of a specific version of Python with specific versions of Python libraries in it.

Let's create a dedicated Python environment in the *DSVM* for this book and name it *pbi*. To do so, you need to perform the following steps:

1. Log into the *DSVM*.

2. Open the command prompt by clicking in the search bar next to the *Windows* sign and type *cmd*.

3. Type the following code to create a conda environment named pbi based on Python 3.7:

```
conda create -n pbi python=3.7
```

The decision to use python 3.7 was based on information obtained from the following Microsoft documentation: `https://docs.microsoft.com/en-us/business-applications-release-notes/october18/intelligence-platform/power-bi-service/pervasive-artificial-intelligence-bi/python-service`. According to the documentation, the Power BI service is compatible with Python 3.x so the current 3.x versions of Python should be compatible.

Now we have a Python environment that we can use for our Python development in this book.

Configuring SQL Server Machine Learning Services in DSVM

Several examples of the book require the use of *SQL Server Machine Learning Services (SSMLS)*. *SSMLS* provides tools that enable you to perform advanced analytics inside the database using R, Python, and some tools that make it easier to work with big data. *SSMLS* also offers some pre-trained models built by Microsoft that you can leverage. The preceding features are not part of the default features, but *SSMLS* is enabled by default in the *DSVM*. If you are not using the *DSVM*, you will have to enable it, and instructions of how to do so can be found here: `https://docs.microsoft.com/en-us/sql/machine-learning/install/sql-machine-learning-services-windows-install?view=sql-server-ver15`. The pre-trained models that will be used in the book can be added to your instance of SQL Server as a post-task installation. Go to this URL to get instructions of how to install the pre-trained models as a post-task installation: `https://docs.microsoft.com/en-us/sql/machine-learning/install/sql-pretrained-models-install?view=sql-server-ver15`.

Installing R packages

Some of the R scripts in this book may contain R packages that are not installed on your machine. Installing packages in R is simple and straightforward. The following code installs a popular R package named *data.table* via the R console:

```
install.packages("<package name>")
```

There will be times when you may want to install multiple packages at once. For instance, you may want to install the R package *data.table* and *dplyr* together. You can accomplish that task by creating a character vector that contains two elements, one for *data.table* and another for *dplyr*, and assign the results to a variable named *pkgs*. Then you would pass that variable to the *install.packages()* function as illustrated here:

```
pkgs <- c("data.table", "dplyr")
install.packages(pkgs)
```

The *R character vector* data type is a one-dimensional array of the character data type. You will learn more about this data types as well as other R data types throughout the book.

When you are creating R visuals in Power BI, you need to be cognizant of which version of the package is being used by the Power BI service. You can get a list of all the available R packages in the Power BI service along with their version at this URL: https://docs.microsoft.com/en-us/power-bi/service-r-packages-support.

The *install.packages()* will download the most recent version of the package from the repository you are using if you are using the distribution of R from *CRAN*. If you are using *Microsoft R Open*, it will install the most recent package based on the *snapshot* date. Both methods may result in a version being installed that is not the same as the one in the service. To download the version of a package that is being used in the service, you first need to get the package version from the page located at the preceding URL, then you need to use the *devtools* package to install it. Here is an example of using the *devtools* package to install *ggplot2 0.9.1* from CRAN:

```
library(devtools)
install_version(
    "ggplot2",
    version = "0.9.1",
    repos = "http://cran.us.r-project.org")
```

Installing Python libraries

There are multiple ways you can install libraries in Python. You will use two methods in this book, *conda* install and *pip* install. Installing libraries in Python is not as easy and straightforward as it is installing packages in R. In this book, you will use the *conda* prompt to install Python libraries. Perform the following steps to install Python a library using *conda*:

1. Go to the *Windows search bar* located on the lower right next to the *Windows* icon.

2. Type the word Anaconda and the *Anaconda Prompt* should appear in the returned list. Click it to launch the *Anaconda Prompt*.

3. Activate the environment that you are using for Power BI by typing the following code in the command prompt:

   ```
   conda activate "<environment name>"
   ```

   It is highly recommended to use an environment for the Python development associated with this book. Instructions of how to create one are located in the next section.

4. Install the package using conda with the following code:

   ```
   conda install <"package name">
   ```

   So, if you were installing pandas, the code would be as follows:

   ```
   conda install pandas
   ```

   If you wanted to install pandas 1.0.4, you would use the following code:

   ```
   conda install pandas=1.0.4
   ```

   Not all packages are available for a *conda* install. Go to the following URL to get a list of packages that can be installed using *conda* in Python 3.6: `https://docs.anaconda.com/anaconda/packages/py3.6_win-64/`. One of the packages that will be used in this book is not available in conda but is available in *PyPI*. The name of the package is *CensusData*. You must use *pip* to install the *CensusData* library as illustrated here:

```
pip install CensusData
```

Configure Power BI in DSVM

There are several configurations that you need to do in the *Power BI Desktop* to enable R and Python. Those steps are covered in detail in the book's code repository in *GitHub*. You will also find instructions on how to create and use *conda* environment in Python. Here is the URL to the book's code repository: https://github.com/Apress/adv-analytics-in-power-bi-w-r-and-python.

Alternative setup

That *DSVM* is the optimal way to go, but it may not be an option for you. If that is the case, you will have to manually install the required software. Here are links to the required software that you need to install:

- Manually Install Power BI: www.microsoft.com/en-us/download/details.aspx?id=58494

- Manually Install R Studio: https://rstudio.com/products/rstudio/download/

- Manually Install Microsoft R Open: https://mran.microsoft.com/download

- Manually Install Anaconda: www.anaconda.com/products/individual

- Manually Install VS Code: https://code.visualstudio.com/download

- Manually Install SQL Server 2019 Developer: www.microsoft.com/en-us/sql-server/sql-server-downloads

- Manually Install SQL Server Machine Learning Services:
  https://docs.microsoft.com/en-us/sql/machine-learning/
  install/sql-machine-learning-services-windows-
  install?view=sql-server-ver15

If you go this route, I highly recommend you use a virtual machine running *Windows Server 2016* or later. Here is the URL to a YouTube video that provides step-by-step instructions on how to install *Windows Server 2019* in *VirtualBox*: www.youtube.com/watch?v=ZjQSuyuNOnA&t=8s.

Adding R packages to SQL Server Machine Learning Services

In Chapter 10 of the book, you will learn how to productionize machine learning models via *SQL Server Machine Learning Services 2019 with R*. When you do so, you need to make sure that the required R packages are loaded in *SSMLS 2019*. Here is a T-SQL script that you can use to see what packages are loaded in your instance of *SSMLS 2019*:

```
EXECUTE sp_execute_external_script
  @language=N'R',
  @script = N'
packagematrix <- installed.packages();
Name <- packagematrix[,1];
Version <- packagematrix[,3];
OutputDataSet <- data.frame(Name, Version);'

WITH RESULT SETS ((PackageName nvarchar(250), PackageVersion nvarchar(max) ))
```

If the package you need is not in the list, then you will need to load it manually. Here are the recommended steps to use to load the R packages in *SSMLS 2019*.

Step 1: Download sqlmlutils on your machine to the Documents folder

The download for *sqlmlutils* can be found at this URL: `https://github.com/Microsoft/sqlmlutils/tree/master/R/dist`.

Download the zip file from this GitHub repo and save it to your *Documents* folder.

Step 2: Run the following code in the command prompt

Open the command prompt as administrator and run the following code:

```
R -e "install.packages('RODBCext', repos='https://cran.microsoft.com')"
R CMD INSTALL %UserProfile%\Documents\sqlmlutils_0.7.1.zip
```

The preceding code will work if you installed *sqlmlutils* in the *Documents* folder under your profile. You will need to change the file path to the appropriate location if *sqlmlutils* was saved in another location.

Step 3: Load the required packages

After you perform Step 2, you will be in the position to load packages to *SSMLS 2019* from an R script in R Studio. Here is a code snippet that loads the *dplyr* package to *SSMLS 2019*:

```
library(sqlmlutils)
connection <- connectionInfo(
  server   = "server",
  database = "database",
  uid      = "username",
  pwd      = "password")

sql_install.packages(connectionString = connection,
pkgs = "dplyr", verbose = TRUE, scope = "PUBLIC")
```

You are not limited to loading multiple packages at once. Let's say you want to load two packages at once, *dplyr* and *data.table*. You could do so by creating a character vector that contains both packages and use it for the *pkgs* argument as shown here:

```
library(sqlmlutils)
connection <- connectionInfo(
   server   = "<server>",
   database = "<database>",
   uid      = "<username>",
   pwd      = "<password>")

pkgList <- c("dplyr","data.table")
sql_install.packages(connectionString = connection,
pkgs = pkgList, verbose = TRUE, scope = "PUBLIC")
```

Add necessary Python libraries to SQL Server Machine Learning Services

Just like with *SQL Server Machine Learning Services 2019 with R*, you need to know how to load Python packages for *SQL Server Machine Learning Services 2019 with Python* in Chapter 10. Here are the steps you should take to load Python libraries in *SSMLS 2019* using *sqlmlutils*.

Step 1: Download sqlmlutils on client machine to the Documents folder

The download for *sqlmlutils* can be found at this URL: https://github.com/Microsoft/sqlmlutils/tree/master/Python/dist.

Download the zip file located in the preceding URL and save it to your *Documents* folder.

Step 2: Open the command prompt and run the following code

```
pip install "pymssql<3.0"
pip install --upgrade --upgrade-strategy only-if-needed c:\temp\sqlmlutils-
0.7.2.zip
```

Step 3: Load the required packages

After you perform Step 2, you will be in the position to load packages to *SSMLS 2019*
from a Python script in *VS Code*. Here is a code snippet that loads the *pandas* package to
SSMLS 2019:

```
import sqlmlutils
connection = sqlmlutils.ConnectionInfo(
    server="<yourserver>", database="<yourdatabase>",
    uid="<username>", pwd="<password>"))
sqlmlutils.SQLPackageManager(connection).install("pandas")
```

On-premises data gateway

The *on-premises data gateway* is a tool that is used to enable secure data movement from
on-prem data sources to the Azure cloud. The *on-premises data gateway* can be run in
two different modes, personal and standard. If you want to deploy the solutions created
in Chapters 3–9 to the Power BI service and run them on a periodical basis, then you will
need to run the *on-premises data gateway* in personal model. As of the writing, R and
Python scripts used in Power Query can only be run via the *on-premises data gateway*
when it is in the personal mode. The downside of using the *on-premises data gateway*
in personal mode is that your solution will not be an enterprise solution. It will be a
solution that only you will have access to because, as the name suggests, the personal
mode is meant for individual use.

However, in Chapter 10, you will learn how you can leverage both R and Python in
an enterprise Power BI solution via *SQL Server Machine Learning Services (SSMLS)*.
When you use *SSMLS 2019*, your R and Python code will be wrapped in a special T-SQL
stored procedure. T-SQL stored procedures can be used in the standard mode of the

on-premises data gateway. In Chapter 10, you will learn the basics of how to refactor the R and Python code you wrote in Chapters 3–9 for the special stored procedure used in *SSMLS 2019*. Note that R visuals do not rely on the *on-premises data gateway* because they are rendered using the instance of R located in the Power BI service.

Resources

This book touches many technologies, so it would be impossible to give a complete coverage to all of them. Realizing that, I decided to give a curated list of resources that you can use to help fill in the missing pieces. I also included a link to the code repository for the book as well as some good general education resources.

Book's code repository

The code for each exercise in this book is contained in the book's code repository at this URL: `https://github.com/Apress/adv-analytics-in-power-bi-w-r-and-python`. The complete R and Python scripts are grouped by chapter and topic in the code repository. Also, either the actual data sources or information on how to acquire the data sources used in the examples is located in the code repository as well.

R resources

Books

- *R for Data Science*: This excellent book properly introduces you to R in a way that is recommended by arguably the most prolific R package creator, Hadley Wickham. It does so by leveraging packages that are a part of *tidyverse*. The book is available for free at this URL: `https://r4ds.had.co.nz/`.

- *An Introduction to Statistical Learning: With Applications in R*: This book is a good introduction to statistical concepts that are important to machine learning via R. The book has been around for a few years and is very popular in the R community. It is often used as the main textbook for undergraduate- and graduate-level courses at many colleges and universities. You can acquire the book for free at this URL: `https://faculty.marshall.usc.edu/gareth-james/ISL/ISLR%20Seventh%20Printing.pdf`.

Websites

- *RStudio*: This website provides a wealth of educational R resources. To get to the educational resources on the site, go to the *Resource* tab in the menu bar and you will see options to access free webinars, cheat sheets, and books. This is also the site where you get the recommended IDE for R, *R Studio*. Here is the URL to the website: `https://rstudio.com/`.

- *The R Graph Gallery*: This site provides examples of visuals created using R complete with the underlying R code. You can use this site to get visualization ideas. Here is the URL to the site: `www.r-graph-gallery.com/`.

- *R Bloggers*: Great website to find community contributed blogs covering various topics about the R programming language. Here is the URL to the blog: `www.r-bloggers.com/`.

Tutorials

- *dplyr tutorial Part 1*: This is the first part of a two-part video series about *dplyr* given by the package main author, Hadley Wickham. The video was made back in 2014, but it is still very relevant. Here is the URL to the tutorial: `www.youtube.com/watch?v=8SGif63VW6E&t=15s`.

- *dplyr tutorial Part 2*: This is the second part of the series mentioned in the previous bullet: `www.youtube.com/watch?v=Ue08LVuk790`.

Python resources

Books

- *Hands-On Machine Learning with Scikit-Learn, Keras, and TensorFlow*: This is an excellent resource that teaches you how to apply *machine learning* and *artificial intelligence* to your data using tools in Python. The book is available for purchases from most book sellers.

- *Python for Data Analysis: Data Wrangling with Pandas, NumPy, and IPython*: This book is written by the original author of the *pandas* library, Wes McKinney. The *pandas* library is arguably the most popular library in Python for data wrangling. The book is available for purchases from most book sellers.

Vlogs, podcasts, and training

- *Data School*: This YouTube channel is authored by Kevin Markham. It provides a wealth of videos covering how to perform tasks in data science using Python libraries such as *pandas, matplotlib, scikit-learn*, and others. He does a great job at breaking down complex concepts in an easy-to-understand way. Here is the URL to this excellent YouTube channel: `www.youtube.com/channel/UCnVzApLJE2ljPZSeQylSEyg`.

- *Google's Python Class*: This is a relatively old, but still relevant, introduction to Python presented by Google. The tutorial does a good job of introducing data structures and other necessary topics required to build a solid foundation to build on. It also includes a very informative section on regular expressions. Here is the URL to the website of this free 2-day Python course: `https://developers.google.com/edu/python`.

- *Talk Python to Me*: This is a very informative podcast covering a wide array of Python topics, many of which are about data analytics. Here is the URL to the podcast's site: `https://talkpython.fm/`.

Websites

- *PEP 8*: One of the best features of Python is how readable Python code is compared to other programming languages. Unlike most languages, Python forces a strict code style to facilitate code readability. The code style is explained in detail in *Pep 8* of Python. You can find out more about *Pep 8* at this URL: `https://pep8.org/`.

Power BI resources

Vlogs

- *Guy in a Cube*: This is hands down, the number one go-to vlog for all things Power BI. They have scores of YouTube videos covering all aspects of Power BI including Power BI administration, Power BI data modeling, and Power BI data visualization. Here is the URL to their YouTube channel: `www.youtube.com/channel/UCFp1vaKzpfvoGaiOvE5VJOw`.

Websites

- *SQLBI*: The creators of this website are the authoritative voice when it comes to the *DAX* expression and query language used in Power BI. Their site contains a wealth of information and tools related to *DAX*. You will find thorough articles covering all things *DAX*, online training, and tools such as *DAX Studio*.

- *Tabular Editor*: The Tabular Editor is a must-have tool when you start doing serious Power BI development. It is an open source tool that will soon be integrated in Power BI. To learn more about this tool, go to this URL which will take you to a website dedicated to this awesome tool: `https://tabulareditor.com/`.

Books

- *The Definitive Guide to DAX*: This is the "go-to" book for those who want to get a thorough understanding of the *DAX* expression and query language used in Power BI. The book is available for purchases from most book sellers.

- *M Is for Data Monkeys*: This book is a good introduction to the *M* functional programming language used in Power Query. The authors do a great job at presenting common data wrangling recipes using *M* in a very easy-to-understand way. It provides a very solid foundation that enables you to approach more complicated *M* topics. The book is available for purchases from most book sellers.

Podcast

- *BIFOCAL*: Great podcast to stay current on all things related to Power BI. You can find the podcast on popular podcast platforms.

General resources

Books

- *Data Science for Business* by Foster Provost and Tom Fawcett: This book introduces data science principles in a program language agnostic way. It is a popular book in the data science community for those that want to learn how to apply data science in business. The book is available for purchases from most book sellers.

Websites

- *Data Science Central*: This is one of the premier sites for data science. Here is the URL to the site: `www.datasciencecentral.com/`.

- *Kaggle*: This site started off as a site for data science competitions, but it has morphed into a site that is so much more. A great site to find data sets that you can use for your machine learning and artificial intelligence education. Here is the URL to the site: `www.kaggle.com/`.

- *ExcelTv*: Microsoft Excel has been, and will continue to be in the foreseeable future, a must-have tool in data science. The guys at *ExcelTv* has created content that covers all aspects of Excel to help make you become very proficient at using the software. The URL to their site is `https://excel.tv/`.

Vlogs and tutorial

- *Excel on Fire*: Oz, the creator of the vlog, is arguably the most passionate guy I ever met when it comes to data wrangling in Excel via *Power Query*. Over the last few years, he has created a wealth of YouTube videos that covers common data wrangling tasks that he has solved using Power Query. The production quality of the videos is top-notch, and his ability to break down hard-to-understand Power Query concepts in an easy-to-understand way is second to none. You can find a link to his YouTube channel at this URL: `www.youtube.com/user/WalrusCandy/featured`.

- *Regular Expression Tutorial* by Corey Schafer: Great introductory but broad tutorial that covers pattern matching for regular expressions. Trust me, after you watch this video, you will not only learn what all those funny characters mean in regex, but you will also gain an appreciation about how powerful and applicable they are! Here is the link to the vlog: `www.youtube.com/watch?v=sa-TUpSx1JA&t=554s`.

Podcasts

- *Analytics on Fire podcast*: This is your one-stop shop for weekly BI/Analytics masterclasses from the top influencers, customers, and thought leaders in the industry. Tune in weekly to get a backstage pass to these engaging discussions which marry education with entertainment. You can find the podcast on popular podcast platforms.

- *Data Skeptic*: This program language agnostic podcast introduces advanced data science topics in an easy-to-understand and entertaining way. He does so via some very interesting examples. You can find this podcast on popular podcast platforms.

- *Freakonomics*: Learning the technical side of data science is only half the battle. It is also very important to develop analytical acumen. This podcast helps you develop that skill. You can find this podcast on popular podcast platforms.

- *SQL Data Partners*: This is a very entertaining and educational podcast covering various topics in the *Microsoft Data Platform*. The hosts of the show are very knowledgeable about all aspects of the *Microsoft Data Platform*. Sometimes their comedic talents come through. Who knows, they may have a future in comedy. Lol. You can find their podcast on popular podcast platforms.

- *Storytelling with Data*: This podcast is presented by Cole Knaflic, and it is a great resource to learn data visualization best practice techniques. She is also the author of a best-selling book under the same name. You can find her podcast on popular podcast platforms.

Chapter summaries

- Chapter 1 – The Grammar of Graphics: Arguably, one of the strengths R has over Python is its data visualization capabilities. The top data visualization package in R is *ggplot2,* and it is based on a concept known as the *grammar of graphics*. This chapter introduces you to the *ggplot2* package and presents a framework on how to properly use the package in Power BI.

- Chapter 2 – Creating R Custom Visuals in Power BI Using ggplot2: One of the biggest benefits of using *ggplot2* is how expressive it allows you to be when creating your visualization. This chapter provides recipes for several visualizations to give you an idea of the type of *R visuals* you can create in Power BI with the *ggplot2* package.

- Chapter 3 – Reading CSV Files: This chapter provides recipes in both R and Python that allows you to dynamically combine *csv* files in a way that would be much harder to do using Power Query.

- Chapter 4 – Reading Excel Files: This chapter provides recipes in both R and Python that allows you to dynamically combine multiple worksheets from several Excel workbooks in a way that would be much harder to do using Power Query.

- Chapter 5 – Reading SQL Server Data: This chapter provides recipes in both R and Python that allows you to load data from SQL Server into the Power BI data model. What makes this recipe beneficial is that it shows you how you can use R and Python to log information about your load in SQL Server.

- Chapter 6 – Reading Data into the Power BI Data Model via an API: This chapter provides recipes in both R and Python that shows the reader how to acquire data for Power BI using APIs, which is much harder to do and, in many instances, impossible to do using Power Query.

- Chapter 7 – Advanced String Manipulation and Pattern Matching: This chapter provides recipes in both R and Python that shows you how to perform advanced string manipulation tasks leveraging regular expressions. Regular expressions are native to R and Python but not available in Power Query.

- Chapter 8 – Calculated Columns Using R and Python: This chapter provides recipes in both R and Python that show you how to create expressions based on advanced mathematical formulas. You will learn the basics of refactoring mathematical formulas, and you will also learn how to use pre-built functions that abstract the complexity of many calculations from you. The chapter uses the *Haversine* formula to make the example.

- Chapter 9 – Applying Machine Learning and AI to Your Power BI Data Models: This chapter covers multiple topics, involving machine learning and artificial intelligence. It starts by showing examples of how to apply custom machine learning models built in R and Python to your Power BI data models. Then it shows how to enhance your Power BI data models with *Microsoft Cognitive Services* without the need of Power BI Premium. This chapter does not limit you to the capabilities offered by Microsoft because it finishes with showing you how to leverage *IBM Watson Natural Language Understanding* to show you how to leverage some text analytics capabilities that are not in *Microsoft Cognitive Services.*

- Chapter 10 – Productionizing Data Science Models and Data Wrangling Scripts: This chapter shows you how to leverage R and Python in enterprise solutions for Power BI. The methods shown focus on free solutions that are available to users who already have the on-prem version of *SQL Server 2017* and later.

Now that we have all the preliminary stuff out the way, we can get into the contents of the book. The top feature that Power BI is known for is its visualization capabilities. Given that, the book will start off by introducing you to how the R programming language can be used to greatly enhance Power BI's visualization capabilities via the *ggplot2* package.

PART I

Creating Custom Data Visualizations Using R

CHAPTER 1

The Grammar of Graphics

Data visualization has always been a very important part of statistics. It gives statisticians a way to share their findings with others in a concise and relatively easy-to-understand way. Given the fact that R is a programming language built by statisticians for statisticians, the R community has made a considerable effort to ensure that R users are able to effectively create visualizations to help them tell their data stories.

Arguably, the most popular R package for data visualization is the *ggplot2* package. This package was created by Hadley Wickham[1] and is based on the concept of the *layered grammar of graphics*. The *layered grammar of graphics* defines a plot in the following way:

- A default data set and a set of mappings from variables to aesthetics

- One or more layers, each composed of a geometric object, a statistical transformation, a position adjustment, and optionally a data set and aesthetic mappings

- One scale for each aesthetic mapping

- A coordinate system

- The faceting specification

The preceding definition is a concise and thorough explanation of the *layered grammar of graphics* from the author. You will learn how to use the preceding concepts to build beautiful visualizations in Power BI using the *ggplot2* package.

---

[1]Wickham, Hadley. *Ggplot2: Elegant Graphics for Data Analysis, 2nd edition.* Houston, TX: Springer, 2016. p. 86.

© Ryan Wade 2020
R. Wade, *Advanced Analytics in Power BI with R and Python*, https://doi.org/10.1007/978-1-4842-5829-3_1

Even though Power BI gives you the ability to create visuals using multiple libraries and packages in both Python and R, this book will focus specifically on the *ggplot2* package along with a few helper packages. The *ggplot2* package stands in a class by itself compared to other options in R and Python. The *layered grammar of graphics* of which *ggplot2* is based on cuts down the time it takes to convert a data visualization idea to code. The package is feature-rich, so a large portion of your data visualization needs can be met with *ggplot2* and its helper packages.

Steps to build an R custom visual in Power BI

This section outlines the workflow that you will use in Power BI to create an R visual. The steps are consistently used regardless of the type of R visual you are creating. Here are the steps.

Step 1: Configure Power BI

You were given information on how to configure R in Power BI in the book's introduction. Make sure you go through the steps outlined in those instructions to ensure Power BI is configured to use R.

Step 2: Drag the "R custom visual" icon to the Power BI canvas

You'll now have access to an *R custom visual* icon. To add an R visualization to Power BI, you first need to drag the R visual icon from the *Visualization Pane* to the *Report Canvas*. The icon for the R visual is a capital *R*, and it is highlighted in the image in Figure 1-1.

Figure 1-1. *The R icon used to create R visuals in Power BI*

The R visual is one of the default visuals available in Power BI. You can size it and position it like any other Power BI visual.

Step 3: Define the data set

Like with any other visual, you need to define the data set that the visual will be based on. You can build up the data set by adding fields from your data model along with optional measures to the *Values* pane. The resulting data set will be exposed to R as an object known as a *data frame*. You can think of a data frame as an *Excel table* but with many more features that facilitate data analysis.

One of the features of a data frame that you need to be aware of when creating an R visual is that the rows in a data frame need to be unique. Given that, R will take steps to make sure the rows are unique when the data frame is created in R. You need to make sure that the data set you are passing to R has unique rows.

Step 4: Develop the visual in your default R IDE

Now that you have a data set defined, you can use that data set to develop the desired R visual in your preferred IDE. One of the things you did when you configured R in Power BI was you identified your preferred R IDE. That feature enables you to develop the code to create your R visual in a feature-rich IDE like *R Studio* instead of the R script editor in Power BI.

The code for your R visual resides in the *R script editor*. Power BI displays it when you select the icon for the R visual in the report canvas. The R script editor looks like that in Figure 1-2.

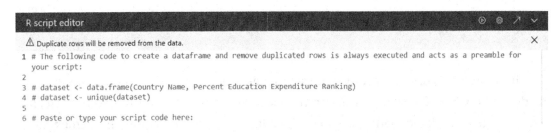

Figure 1-2. *The R script editor in Power BI*

As you can see, this is not an ideal place to develop your R code. There is no intelli-sense, no console, and the amount of real estate you have to develop the code is limited. Fortunately, Power BI gives you the ability to port the data set that you defined for your custom R visual to the R IDE you selected. The recommended IDE for R is *R Studio*, and it will be the one used here if you followed the instructions outlined in the book's introduction.

To port the data set for your visual for use by R in *R Studio*, select the R visual in the report canvas, then click the arrow that is pointing 45 degrees to the right. That action will launch *R Studio* and will create the beginnings of an R script that includes code that produces a data frame named *dataset* based on the data passed to the visual in Power BI. That action also includes any R code associated with the visual in the script. Instead of Microsoft trying to re-invent the wheel, they decided to give you the ability to do your development in an IDE like *R Studio*. Given that, I highly recommend that you do all code development in *R Studio* instead of the tiny R editor that is available to you in Power BI.

Step 5: Use the following template to develop your visual

The following template in Listing 1-1 should be used for the R script when developing R visuals in Power BI:

```r
if (<"data test">) {

    #<"Code for R Visual">

} else {

    plot.new()
    title(main = "<predefined message>")

}
```

This template is used to test the data set to make sure it meets the requirements for the desired R visual. If it does, it will commence to create the visualization; otherwise, it will generate a blank plot with a message. The blank chart with the predefined message is created using the following code:

```r
plot.new()
title(main = " <predefined message> ")
```

The first line creates a new blank chart, and the second line adds a title to the chart containing the message that you want to display to the report user.

Step 6: Make the script functional

Once you get the R script functional, copy the complete script to the R editor. Include all the code except the line that creates the data frame named *dataset*. The *dataset* data frame is automatically generated in Power BI. Note that the resulting R visual in Power BI is totally responsive to outside filters and slicers just like other visuals. You will see several examples in the next chapter. One thing you lose, however, is bidirectional filtering. Outside visuals will propagate to your R visual in Power BI, but you can't filter other Power BI chart items directly from an R visual.

Recommended steps to create an R visual using ggplot2

The previous section illustrated the overall template you should use when defining your overall R script for your R visual. It included code to test your data set as well as code for error handling. In this section, we will focus specifically on the recommended template for R visuals after you have tested the data set. Specifically, we are talking about the code that goes where you see #<"Code for R Visual"> in the following listing.

Listing 1-1. R visual template for Power BI

```
if (<"data test">) {

    #<"Code for R Visual">

} else {

    plot.new()
    title(main = "<predefined message>")

}
```

The steps of developing the template will be described by developing the chart pictured in Figure 1-3.

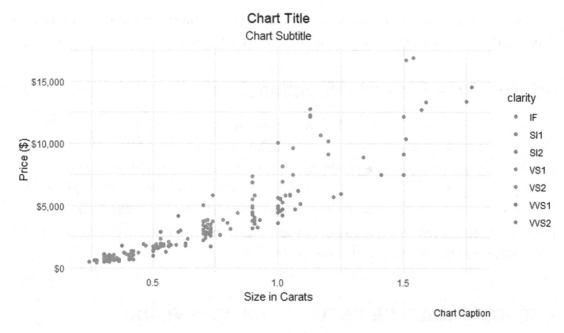

Figure 1-3. *Visual used to describe the template to use when developing R visuals in Power BI*

The preceding R visual uses the diamonds data set to plot a sample of diamonds with a color of *D* to show diamond prices relative to carat size. It is a good example to use because it provides a relatively easy way to illustrate each step in the template when developing an R visual. Let's begin with Step 1.

Step 1: Load the required packages that you will need for your script

For this example, you will use the following code in Listing 1-2 to load the two packages needed for the script.

Listing 1-2. Import statements

```
library(tidyverse)
library(scales)
```

You use *ggplot2* and *dplyr* from the *tidyverse* metapackage.

The *tidyverse* metapackage is a collection of packages that work together to make doing data science in R easier. The *tidyverse* metapackage gives you the ability to import the core packages in one call by loading *tidyverse* instead of having to load each of the core packages individually.

ggplot2 will be used to create the visualization, and *dplyr* will be used to perform some data wrangling tasks. The *scales* packages will be used to help with some number formatting tasks.

Step 2: Make any required adjustments to the data set

Often adjustments need to be made to the data frame before it is passed to *ggplot2*. The code used to make the adjustments to the data frame in this example is in Listing 1-3.

Listing 1-3. Data prep for visual

```
plot.data <-
    diamonds %>%
    filter(color == "D") %>%
    sample_n(200)
```

The data set passed to R contains records that are not needed by the visual. So, you need to use the *dplyr* package to remove unneeded data. You want to create a visualization based on the color type *D*. To get a data frame with the desired color type, you need to filter the original *diamonds* data frame to only include those colors. You do that in the preceding third line.

That filtering leaves you with only diamonds with a *D* color. The total number of records in the resulting data frame is 6775 records which is high for a scatter plot visual. Plotting too many observations in a scatter plot visual can result in a problem known as over-plotting. To minimize over-plotting, you can take a random sample of the data and plot the sampled data instead. The sampled data will have a similar distribution as the underlying data which will cause it to plot similarly but minimize data overlapping. The preceding code does an inline sample via the *sample_n()* function from *dplyr*. The result is a random sample with 200 records that plots much better.

Step 3: Initiate the creation of the visualization with the ggplot() function

You initiate the creation of a chart in *ggplot2* with the *ggplot()* function. When you initiate the *ggplot()* function, you pass it parameters that you want to make available to all layers of your chart. In this example, you set the *plot.data* data frame to the data argument and the *caret* field and *price* field to the x and y argument of the *aes()* function as shown in Listing 1-4.

Listing 1-4. Initiating the ggplot() function

```
ggplot(data = plot.data, aes(x = carat, y = price))
```

It is important to know what the *aes()* function is and how it is used. The *aes()* function describes how to map your data to a visual attribute. Commonly used aesthetics are

- The x coordinate and the y coordinate of the geometric shape you are plotting

- The color of the geometric shape you are plotting

- The size of the geometric shape you are plotting

- The fill of the geometric shape you are plotting

The preceding *aes()* function is used to describe how to map the x and y coordinates of the geometric shape that will be plotted. The *aes()* function can be defined in the *ggplot()* function or in a *geom* function that you will learn about in the next step. If it is defined in the *ggplot()* function, then it will be available to all layers of the chart, but if it is defined in the *geom* function, then it will only be available in the layer it was defined. You will see examples of both cases in the next chapter.

Step 4: Add desired geom(s)

You can do so using the following code in Listing 1-5.

Listing 1-5. Adding a geom

```
ggplot(plot.data, aes(x = carat, y = price)) +
geom_point()
```

When you initiated the *ggplot()* function in the previous step, the visual did not have any layers so there was nothing to visualize. The mechanisms used to add layers to a *ggplot* visual are *geoms*. The geom used to create scatter plots in *ggplot* is *geom_point()*. In the preceding code, *geom_point()* inherits from the *ggplot()* function that preceded it so you should notice that the data source and the aesthetics needed to define the x and y coordinates did not need to be explicitly defined. The resulting chart is shown in Figure 1-4.

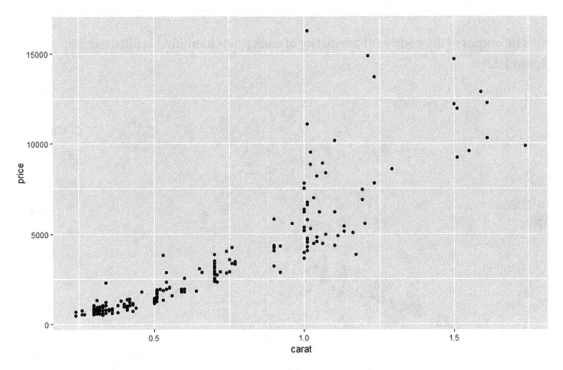

Figure 1-4. *Basic scatter plot*

You want to add an additional layer of detail. You want to identify the *clarity* of each point in the scatter plot. You do so by setting the aesthetic for *color* to *clarity* in the *geom_ point()* geom as illustrated in the code in Listing 1-6.

Listing 1-6. Adding an aesthetic for color

```
ggplot(plot.data, aes(x = carat, y = price)) +
geom_point(aes(color = clarity))
```

You set this aesthetic in the *geom_point()* geom because that aesthetic is only needed in that layer.

It is important to remember that when you set data sources or aesthetics in your ggplot() function, they will be available to all layers in your visual, but when you set them in an individual geom, it will only be available in the layer associated with the geom.

The output of the code with the clarity of each point identified is illustrated in Figure 1-5.

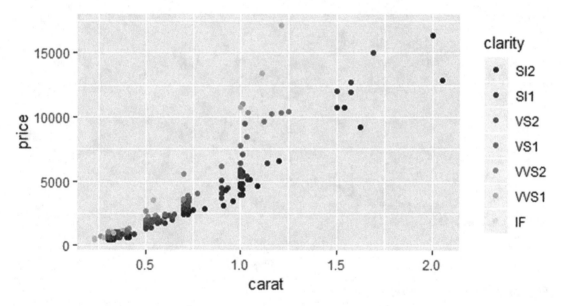

Figure 1-5. *Basic scatter plot*

Step 5: Define your titles, subtitles, and caption

Here is code you can use to add a title to your visual in Listing 1-7.

Listing 1-7. Adding the titles and caption

```
ggplot(plot.data, aes(x = carat, y = price)) +
geom_point(aes(color = clarity))+
labs(
    title = "Chart Title",
    subtitle = "Chart Subtitle ",
    caption = "Chart Caption"
)
```

ggplot2 makes it easy to add a title, subtitle, and caption to your chart via the *labs()* function. The *labs()* function name is short for labels. The titles of your visual can be based on a static string or a dynamically generated string. You will see examples of titles based on dynamically generated strings in the next chapter. Figure 1-6 shows what the chart would look like with the titles added.

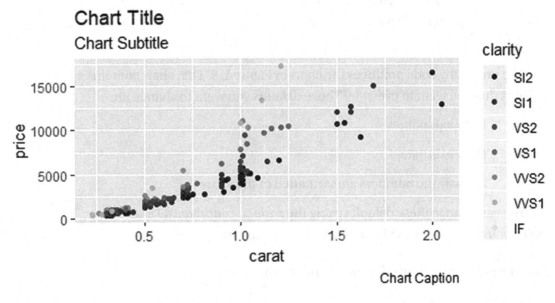

Figure 1-6. *Chart with the titles added*

Step 6: Make any necessary changes to the x and y axis

The defaults labeling of the x and y coordinates provided by *ggplot2* are often sufficient and don't require any change. But if you are not happy with the defaults, you can easily change them using a *scale* function.

Let's take a moment to look at the *ggplot2* code you built up to this point in Listing 1-8.

Listing 1-8. Code to modify the x and y axis

```
library(tidyverse)
library(scales)

plot.data <-
    diamonds %>%
    filter(color == "D") %>%
    sample_n(200)

ggplot(plot.data, aes(carat, price)) +
    geom_point(aes(color = clarity)) +
    labs(
        title = "Chart Title",
        subtitle = "Chart Subtitle",
        caption = "Chart Caption"
    )
```

The preceding code produces the chart in Figure 1-6. That chart contains a few defaults that you want to override. Those defaults you want to change are

- The x axis names

- The y axis names

- The way the numbers are formatted in the y axis

You can change these defaults using the *scale_x_continuous()* and *scale_y_continuous()* functions as illustrated in the code in Listing 1-9.

Listing 1-9. Code to modify the x and y axis

```
library(tidyverse)
library(scales)

plot.data <-
    diamonds %>%
    filter(color == "D") %>%
    sample_n(200)
```

```
ggplot(plot.data, aes(x = carat, y = price)) +
    geom_point(aes(color = clarity)) +
    labs(
        title = "Chart Title",
        subtitle = "Chart Subtitle",
        caption = "Chart Caption"
    ) +
    scale_x_continuous(name = "Size in Carats") +
    scale_y_continuous(
        name = "Price ($)",
        labels = dollar_format()
    )
```

The result you receive after executing the preceding code is illustrated in Figure 1-7.

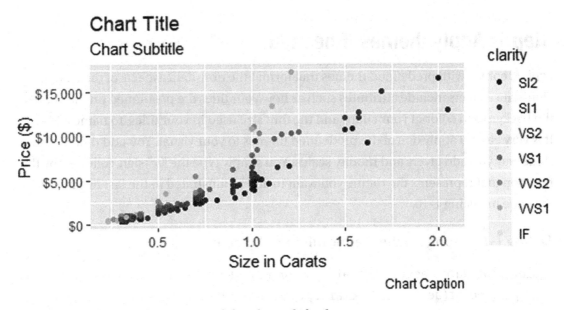

Figure 1-7. *Overriding some of the chart defaults*

As you can see, the *scale_x_continuous()* and *scale_y_continuous()* functions make it relatively easy to make the desired changes.

Notice you add scales to your visual in the same fashion that you used to add layers to your visual. It is important to note that the scale functions are used to override the default scales, but they do not add layers to your chart.

Because both the x and y axis are based on continuous data, the continuous version of that scale for the x and y axis is used. The only thing you want to change for your x axis is the name, and you see that is easily accomplished using the *name* argument. For the y axis, you are not only changing the name, but you are also changing the way the numbers are formatted on the axis. The name change is accomplished using the *name* argument. The formatting is done via the *labels* argument. The *dollar_format()* function from the *scales* package is used to format the numbers to a dollar format. The *dollar_format()* function inherits the y value from the *ggplot()* function that was defined in the first line of the script so it does not need to be explicitly defined in the function.

Step 7: Apply themes if needed

ggplot2 comes with predefined themes that format the non-data aspects of the charts. Non-data aspects include attributes such as how your titles are positioned on the chart, the background color of your chart, and the font size used in your titles, to name a few. It is very easy to apply one of the predefined themes to your visual. You add themes the same way you add layers and modify scales. You simply type the "+" sign, followed by the function that represents the theme you want to add as illustrated in the last two lines in the code in Listing 1-10.

Listing 1-10. Applying the theme_minimal() theme

```
ggplot(plot.data, aes(x = carat, y = price)) +
    geom_point(aes(color = clarity)) +
    labs(
        title = "Chart Title",
        subtitle = "Chart Subtitle",
        caption = "Chart Caption"
    ) +
```

```
scale_x_continuous(name = "Size in Carats") +
scale_y_continuous(
    name = "Price ($)",
    labels = dollar_format()
) +
theme_minimal()
```

The resulting visual is shown in Figure 1-8.

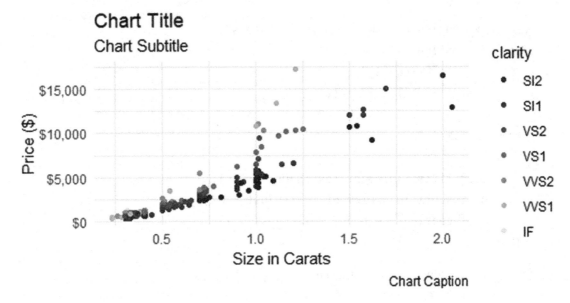

Figure 1-8. *The theme defaults based on the R script*

You used the *theme_minimal()* theme in the preceding visual. It is considered a good practice in data to only add elements to your visual if it is absolutely necessary. White space in your visual is considered a good thing. The *theme_minimal()* theme is based on that principle. A list of the themes that are available in *ggplot2* is covered in a subsequent section titled "Themes built into ggplot2."

Step 8: Use the theme() function to change any specific non-data elements

In the event that you are not totally happy with the results of the theme you applied to your visual, you can tweak the results even further using the *theme()* function. The code in Listing 1-11 uses the *theme()* function to format the titles.

Listing 1-11. Using the theme() function to change non-data elements

```
ggplot(plot.data, aes(carat, price)) +
    geom_point(aes(color = clarity)) +
    labs(
        title = "Chart Title",
        subtitle = "Chart Subtitle",
        caption = "Chart Caption"
    ) +
    scale_x_continuous(name = "Size in Carats") +
    scale_y_continuous(
        name = "Price ($)",
        labels = dollar_format()
    ) +
    theme_minimal() +
    theme(
        plot.title = element_text(hjust = 0.5),
        plot.subtitle = element_text(hjust = 0.5),
        plot.caption = element_text(hjust = 1)
    )
```

The code produces the following visual in Figure 1-9.

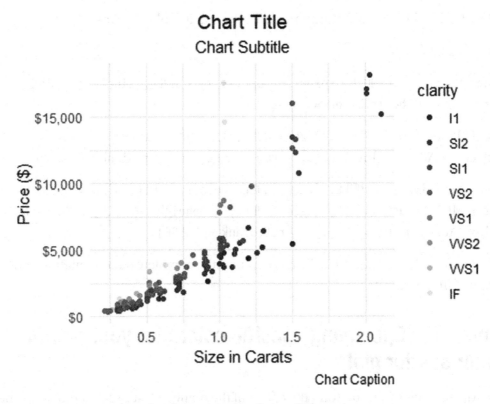

Figure 1-9. *Re-positioned titles using the theme() function*

The theme function gives you access to over 80 chart elements that you can modify. The majority of those elements have an element function associated with it that is used to make the modification.

In the preceding code, we modified the position of the chart's title, the chart's subtitle, and the chart's caption. As you can see in the code, the element associated with the chart title is *plot.title*, the element associated with the chart's subtitle is *plot.subtitle*, and the element associated with the chart's caption is *plot.caption*. All three elements are modified using the *element_text()* function. The chart's title and subtitle were centered using the *hjust* argument in the *element_text()* function. The *hjust* argument represents the horizontal position of the text, and it accepts a value between 0 and 1. If you set *hjust* to 0, it will left justify the text; if you set *hjust* to 0.5, it will center the text; and if you set *hjust* to 1, it will right justify the text.

Listing 1-12 shows the element functions along with their available options.

Listing 1-12. The element functions with their available options

```
element_blank()

element_rect(fill = NULL, colour = NULL, size = NULL, linetype = NULL,
color = NULL, inherit.blank = FALSE)

element_line(colour = NULL, size = NULL, linetype = NULL,
  lineend = NULL, color = NULL, arrow = NULL, inherit.blank = FALSE)

element_text(family = NULL, face = NULL, colour = NULL, size = NULL,
hjust = NULL, vjust = NULL, angle = NULL, lineheight = NULL, color = NULL,
margin = NULL, debug = NULL, inherit.blank = FALSE)
```

This URL, `https://bit.ly/34EaWtr`, contains the complete documentation about the *theme()* function.

Bonus step: Specifying specific colors for your points in your scatter plot

There will be times when you want the colors of the points in your scatter plot to be the same color. Let's see what happens if we try to set the color aesthetic to a specific color rather than a field in the data frame. The script in Listing 1-13 sets the color to *blue* in the *aes()* function.

Listing 1-13. Setting the color via the aes() function

```
library(tidyverse)
library(scales)

plot.data <-
    diamonds %>%
    filter(color == "D") %>%
    sample_n(200)

ggplot(plot.data, aes(carat, price)) +
  geom_point(aes(color = "blue")) +
  labs(
    title = "Chart Title",
```

```
  subtitle = "Chart Subtitle",
  caption = "Chart Caption"
) +
scale_x_continuous(name = "Size in Carats") +
scale_y_continuous(
  name = "Price ($)",
  labels = dollar_format()
) +
theme_minimal() +
theme(
  plot.title = element_text(hjust = 0.5),
  plot.subtitle = element_text(hjust = 0.5),
  plot.caption = element_text(hjust = 1)
)
```

It produces the following chart in Figure 1-10.

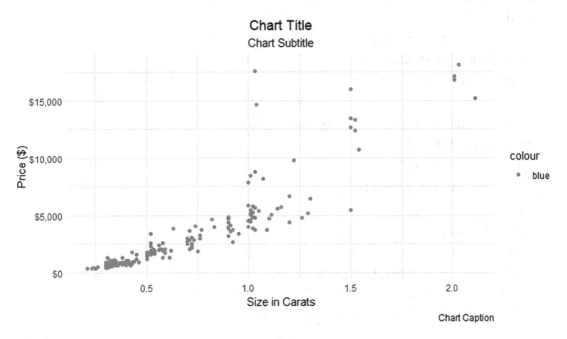

Figure 1-10. *Setting color using the aes() function*

Note that you did not get the expected results. Instead of getting a scattered plot chart with the points colored blue, you got a scattered plot chart with the points colored what appears to be red. That is because the *aes()* function scaled the color. You need to define the color outside of the *aes()* function if you want it to be a specific color as done in the code in Listing 1-14 on line 10.

Listing 1-14. Setting the color outside the aes() function

```
library(tidyverse)
library(scales)

plot.data <-
    diamonds %>%
    filter(color == "D") %>%
    sample_n(200)

ggplot(plot.data, aes(carat, price)) +
  geom_point(color = "blue") +
  labs(
    title = "Chart Title",
    subtitle = "Chart Subtitle",
    caption = "Chart Caption"
  ) +
  scale_x_continuous(name = "Size in Carats") +
  scale_y_continuous(
    name = "Price ($)",
    labels = dollar_format()
  ) +
  theme_minimal() +
  theme(
    plot.title = element_text(hjust = 0.5),
    plot.subtitle = element_text(hjust = 0.5),
    plot.caption = element_text(hjust = 1)
  )
```

It produces the following chart in Figure 1-11 with the desired results.

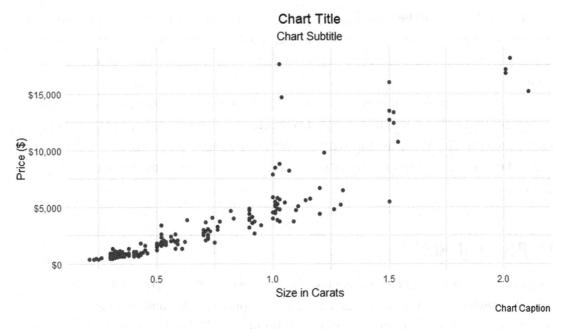

Figure 1-11. *Specifying color outside the aes() function*

The importance of having "tidy" data

For those of you that are DAX developers, you are aware of the importance of having your BI data model formatted in a nice star schema structure. Doing so makes developing DAX measures much easier. Measures that are simple to develop in a well-formed *star schema* data model can be unnecessarily complex in a BI data model that is not shaped in that format.

A similar phenomenon exists when building visualizations or doing data analysis in R. When you format your data in what Hadley Wickham calls a *tidy* format, it makes data analysis tasks and data visualization much easier to do.

A data set that is in a *tidy* format is a data set where each record represents a single observation, and each column represents a single variable. It is similar to the third normal form in databases.

If for some reason the data that you pass to your R visual in Power BI is not in a tidy format, don't panic. Luckily, you have packages such as *dplyr* and *tidyr* that you can use in your R visual script that will help you shape your data into the format it needs to be in.

The *tidy* data concept is very important in data analysis and data visualization. Its importance is not limited to R. A complete coverage is beyond the scope of this book. Fortunately, Hadley Wickham has written a nice white paper that you can read to learn more about it. Here is the URL to that paper, `https://vita.had.co.nz/papers/tidy-data.pdf`.

Popular geoms

As mentioned earlier, *ggplot2* is based on the *layered grammar of graphics*. Layers are added to *ggplot2* visuals through the use of *geom* functions. This feature of *ggplot2* enables you to create beautiful visuals with minimum code that is relatively easy to write. It makes expressing your ideas with code much easier to do compared to creating data visualizations in other programming languages. There are many *geoms* available to you in *ggplot2*, but here is a bulleted list of some popular geoms complete with code snippets to illustrate them:

- **geom_bar()**

  The *geom_bar()* geom is used to create what are typically called column charts and horizontal bar charts. This section gives an example of how to use the *geom_bar()* geom to create a column chart using the code in Listing 1-15.

Listing 1-15. geom_bar() example

```
library(tidyverse)

plot.data <-
  data.frame(Titanic) %>%
  group_by(Class) %>%
  summarize(`Total Freq` = sum(Freq))
```

```
ggplot(plot.data, aes(x = Class, y = `Total Freq`)) +
  geom_bar(stat = "identity") +
  labs(title = "geom_bar") +
  theme_minimal()
```

The visualization produced by the preceding code is shown in Figure 1-12.

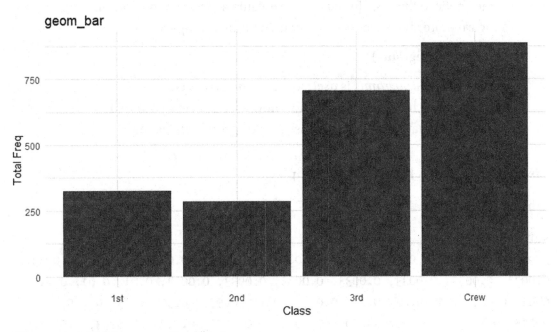

Figure 1-12. *The geom_bar() geom*

The first line of code in the preceding script loads the *tidyverse* package. The next block of code creates the data set needed for the *geom_bar()* visual. The techniques used to wrangle the data in that section will be covered in future parts of the book.

The next block code is the code that is actually creating the visualization. The first two lines are all that is needed to create the column chart. The last two lines are used to change the cosmetics of the chart. They add a title to the chart and changed the chart's theme.

Note that the *geom_bar()* only needed one argument to be defined, and that is the *stat* argument. The other required arguments (the data set, x coordinate, and y coordinate) were inherited from the *ggplot()* function. Setting the stat argument to "*identity*" tells *ggplot2* that you want the height of your columns to be based on the values of the columns that were assigned to the y aesthetic in the *aes()* function. The default is to use count. You will learn more about the *stat* argument in future examples.

- **geom_histogram()**

The *geom_histogram()* is used to create histogram chart. Histogram charts are used to show how your data is distributed. The code that we will use to illustrate this geom is in Listing 1-16.

Listing 1-16. geom_histogram() example

```
library(tidyverse)

set.seed(50)
probs <- c(0.0033, 0.0033, 0.0033, 0.0033, 0.0033, 0.0033, 0.0033, 0.0033,
0.0033, 0.0033, 0.0033, 0.0033, 0.0033, 0.0033, 0.0033, 0.01, 0.01, 0.01,
0.01, 0.01, 0.01, 0.01, 0.01, 0.01, 0.01, 0.015, 0.015, 0.015, 0.015,
0.015, 0.015, 0.015, 0.015, 0.015, 0.015, 0.0133, 0.0133, 0.0133, 0.0133,
0.0133, 0.0133, 0.0133, 0.0133, 0.0133, 0.0133, 0.0133, 0.0133, 0.0133,
0.0133, 0.0133, 0.0133, 0.0133, 0.0133, 0.0133, 0.0133, 0.0133, 0.0133,
0.0133, 0.0133, 0.0133, 0.0133, 0.0133, 0.0133, 0.0133, 0.0133, 0.017,
0.015, 0.015, 0.015, 0.015, 0.015, 0.015, 0.015, 0.015, 0.015, 0.01, 0.01,
0.01, 0.01, 0.01, 0.01, 0.01, 0.01, 0.01, 0.01, 0.0033, 0.0033, 0.0033,
0.0033, 0.0033, 0.0033, 0.0033, 0.0033, 0.0033, 0.0033, 0.0033, 0.0033,
0.0033, 0.0033, 0.0033)

weights <- 265:364

names <- c("Liam  Galles", "Noah  Raymond", "William  Hammontree",
"James  Zaremba", "Oliver  Zurcher", "Benjamin  Hilker", "Elijah  Loken",
"Lucas  Lewter", "Mason  Straus", "Logan  Work", "Alexander  Jarret",
"Ethan  Wey", "Jacob  Adolphsen", "Michael  Solt", "Daniel  Welcome",
```

"Henry Portman", "Jackson Tichenor", "Sebastian Free", "Aiden Papp",
"Matthew Lenzi", "Samuel Rinaldo", "David Goines", "Joseph Asuncion",
"Carter Philhower", "Owen Freeborn", "Wyatt Ice",
"John Mcguckin", "Jack Soden", "Luke Humfeld", "Jayden Natera",
"Dylan Galles", "Grayson Raymond", "Levi Hammontree", "Isaac Zaremba",
"Gabriel Zurcher", "Julian Hilker", "Mateo Loken", "Anthony Lewter",
"Jaxon Straus", "Lincoln Work", "Joshua Jarret", "Christopher Wey",
"Andrew Adolphsen", "Theodore Solt", "Caleb Welcome",
"Ryan Portman", "Asher Tichenor", "Nathan Free", "Thomas Papp",
"Leo Lenzi", "Isaiah Rinaldo", "Charles Goines", "Josiah Asuncion",
"Hudson Philhower", "Christian Freeborn", "Hunter Ice",
"Connor Mcguckin", "Eli Soden", "Ezra Humfeld", "Aaron Natera",
"Landon Galles", "Adrian Raymond", "Jonathan Hammontree",
"Nolan Zaremba", "Jeremiah Zurcher", "Easton Hilker", "Elias Loken",
"Colton Lewter", "Cameron Straus", "Carson Work", "Robert Jarret",
"Angel Wey", "Maverick Adolphsen", "Nicholas Solt", "Dominic Welcome",
"Jaxson Portman", "Greyson Tichenor", "Adam Free", "Ian Papp",
"Austin Lenzi", "Santiago Rinaldo", "Jordan Goines", "Cooper Asuncion",
"Brayden Philhower", "Roman Freeborn", "Evan Ice", "Ezekiel Mcguckin",
"Xavier Soden", "Jose Humfeld", "Jace Natera", "Jameson Adolphsen",
"Leonardo Solt", "Bryson Welcome", "Axel Portman", "Everett Tichenor",
"Parker Free", "Kayden Papp", "Miles Lenzi", "Sawyer Rinaldo",
"Jason Goines")

```r
linemen_weights <- sample(weights, 100, replace = TRUE,
prob= probs)

plot.data <- data.frame(names, linemen_weights)

ggplot(plot.data, aes(linemen_weights)) +
  geom_histogram(binwidth = 20) +
  labs(title = "geom_histogram") +
  theme_minimal()
```

The preceding code produces the visualization in Figure 1-13 which is illustrated as follows.

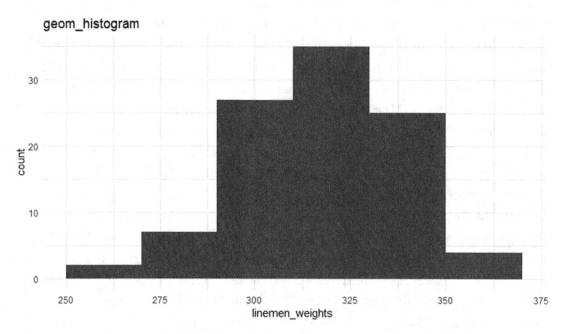

Figure 1-13. *The geom_history() geom*

As in the previous example, the first part of the code builds out the data set that will be used for the visual so you can reproduce the same visual in your environment. It produces a fictitious data set of American football offensive linemen at the collegiate level. We will just focus on the last four lines of code which produces the visual.

As you might have guessed, the basis of the histogram visual is produced using the *geom_histogram()* geom. Note that it uses one argument, and that is *binwidth* because the other required arguments were inherited from the *ggplot()* function. This argument enables you to define the width of the bins. It is recommended you play with this argument because the default settings normally do not create the optimum number of bins. Note that the basis of the histogram was created with just the two following code lines as illustrated in the following code:

```
ggplot(plot.data, aes(linemen_weights)) +
  geom_histogram(binwidth = 20)
```

The subsequent code after that point were for cosmetic features.

- **geom_line()**

  The *geom_line()* geom is used to create line charts. Line charts are great for showing trend over time. The following code in Listing 1-17 will be used to generate the line chart example.

Listing 1-17. geom_line() example

```
library(tidyverse)
library(lubridate)

players <- c("Kobe Bryant", "Pau Gasol", "Lamar Odom")
plot.data <- lakers %>%
  filter(
    result == "made" & team == "LAL" & player %in% players
  ) %>%
  mutate(date = ymd(date)) %>%
  group_by(player, date) %>%
  summarize(points = sum(points))

ggplot(plot.data, aes(x=date,y=points,color=player)) +
  geom_line() +
  labs(title = "geom_line") +
  theme_minimal()
```

The output of the preceding code is shown in Figure 1-14.

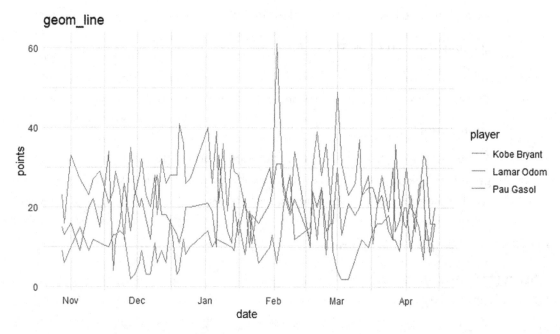

Figure 1-14. *The geom_line() geom*

As in the previous example, the first part of the code builds the data set and gives you everything you need to recreate the visual in your environment. This visualization is using the *lakers* data set that comes with the *lubridate* package. The *lakers* data set contains play-by-play information from the 2008 season.

Note how the color aesthetic is used to create multiple lines. *ggplot2* creates a line based on each unique item in the field specified in the color aesthetic. Just like in the previous examples, only two lines were needed to create the basis of the visual and those were the line where the *ggplot()* function was initiated and the next line after that which defined the *geom* layer.

- **geom_point and geom_smooth**

    This example illustrates how to create a chart in *ggplot2* that has multiple layers. The example shows how to add a regression line to a scatter plot. *ggplot2* enables you to overlay your scatter plot with different types of trend lines. It is not limited to linear regression lines. *ggplot2* also allows you to add confidence intervals to your line at a level of your choice. As of this writing, those features are not available in native Power BI.

    The following code in Listing 1-18 is used to illustrate how to create a scatter plot and overlay it with a regression line.

Listing 1-18. geom_point() and geom_smooth() example

```
library(tidyverse)
set.seed(1)
plot.data <-
  diamonds %>%
  filter(color == "D") %>%
  sample_n(200)

ggplot(plot.data, aes(x=carat, y=price)) +
  geom_point() +
  geom_smooth(method ="lm") +
  labs(title = "geom_point & geom_smooth") +
  theme_minimal()
```

Figure 1-15 shows what the code produces.

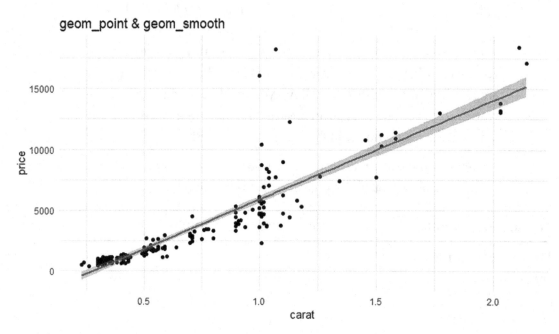

Figure 1-15. *The geom_point() and geom_smooth() geom*

> The preceding chart is based on multiple layers, the *geom_point()*
> and *geom_smooth()*. The *geom_point()* geom is used to create
> the scatter plot, and the *geom_smooth()* geom is used to add a
> regression line complete with a confidence interval. The default
> for the confidence interval is 95%. You can change it by modifying
> the *se* argument. Setting *method* argument to *lm* is what causes
> the *geom_smooth()* geom to produce a regression line.

Controlling aesthetics with scales

You learned earlier that the mapping of data to aesthetics is described using the *aes()*
function. The underlying process, however, is controlled by the *scale* functions. In this
section, you will learn how that happens.

Let's start with a background of what *scales* are. A *scale* is needed for every aesthetic in a chart (x coordinate, y coordinate, color, etc.). It determines how the data will be "scaled" in your visual. You don't need to explicitly define them. If you don't define them, *ggplot2* will go with the defaults. Let's use the code snippet in Listing 1-19 as an example.

Listing 1-19. ggplot2 code minus the underlying helper code

```
ggplot(plot.data, aes(x = carat, y = price)) +
geom_point(aes(color = clarity))
```

In this example, there are no scales being explicitly defined. The preceding code is translated to the following code in Listing 1-20 by *ggplot2*.

Listing 1-20. ggplot2 code with helper code that defines the scales

```
ggplot(plot.data, aes(x = carat, y = price)) +
geom_point(aes(color = clarity)) +
scale_x_continuous() +
scale_y_continuous() +
scale_color_discrete()
```

ggplot2 uses the default values for each of those three scales.

The astute reader probably recognized a pattern in the naming convention of scales. The scale functions use a three-part naming convention. Scale functions are prefaced with the word "scale", followed by the name of the aesthetic it is a scale for, and ending with the type of scale it is. These three parts are delimited with a "_".

So, if you look at the first two scales in the preceding script using the naming convention just outlined, you notice that they are for the x and y aesthetic, respectively. They both use the *continuous* scale type because the data mapped to the x and y aesthetic are continuous numbers. The last scale is for the *color* aesthetic. The data mapped to this aesthetic is the *clarity* field. That field contains a list of discrete values which is the reason why the type of scale is discrete.

If you are cool with what *ggplot2* produces, then you don't need to define the scales. You can go with the defaults. But if you want to override what *ggplot2* produces, you need to do so using the appropriate scale.

Recall the following code from Listing 1-9 that produced Figure 1-8:

```
library(tidyverse)
library(scales)

plot.data <-
    diamonds %>%
    filter(color == "D") %>%
    sample_n(200)

ggplot(plot.data, aes(x = carat, y = price)) +
    geom_point(aes(color = clarity)) +
    labs(
        title = "Chart Title",
        subtitle = "Chart Subtitle",
        caption = "Chart Caption"
    ) +
    scale_x_continuous(name = "Size in Carats") +
    scale_y_continuous(
        name = "Price ($)",
        labels = dollar_format()
    )
```

Note that the aesthetic name is changed on the x aesthetic using the *scale_x_continuous()* and the label formatting and aesthetic name is changed on the y aesthetic using *scale_y_continuous()*. You will see more examples of ways you can modify your aesthetics with scales in the next chapter.

Themes built into ggplot2

ggplot2 comes pre-packaged with a list of themes that makes it easy to change the display of your non-data elements in your chart. Here is a list of available themes as described by the package author, Hadley Wickham:[2]

- *theme_bw()*: A variation on *theme_gray()* that uses a white background and thin gray grid lines

---

[2]Wickham, Hadley. *Ggplot2: Elegant Graphics for Data Analysis*, 2nd edition. Houston, TX: Springer, 2016. pp. 172–173.

- *theme_linedraw()*: A theme with only black lines of various widths on white backgrounds, reminiscent of a linedrawing

- *theme_light()*: Similar to *theme_linedraw()* but with light gray lines and axes, to direct more attention toward the data

- *theme_dark()*: The dark cousin of *theme_light()*, with similar line sizes but a dark background. Useful to make thin colored lines pop out

- *theme_minimal()*: A minimalistic theme with no background annotations

- *theme_classic()*: A classic-looking theme, with x and y axis lines and no gridlines

- *theme_void()*: A completely empty theme

Using R visuals in the Power BI service

When you deploy an R visual to the Power BI service, you must make sure that you are using a version of the package that is available in the service. The following URL contains a list of available packages on the server along with their version: `https:// docs.microsoft.com/en-us/power-bi/connect-data/service-r-packages-support`.

Helper packages for ggplot2

ggplot2 is a feature-rich package, and it contains the tools needed to handle most of your data visualization needs. But there will be times when you will need to do things such as wrangle the data set that Power BI passed to R before you create the visualization, format the data in a way that is not possible using the features available in base R or *ggplot2*, or use themes that are not part of the *ggplot2* package.

Fortunately, you are not limited to the *ggplot2* package when you develop your R visual. Here are a few packages that are great complements to *ggplot2* in Power BI. This list is by no means exhaustive:

- *tidyverse*: A metapackage that contains a collection of popular R packages that share the same framework. It includes packages such as *ggplot* and *dplyr*. The *tidyverse* collection of packages facilitates doing data science in R. More information about *tidyverse* can be found at this URL: `www.tidyverse.org/`.

- *ggthemes*: Is a helper package for *ggplot2* that contains additional scales, geoms, and themes. Two of the themes that are worth noting are *theme_few()* and *theme_tufte()*. These two themes are based on best practices of two very popular data visualization experts, Stephen Few and Edward Tufte. More information can be found about the *ggthemes* package at this URL: `https://cran.r-project.org/web/packages/ggthemes/ggthemes.pdf`.

- *scales*: The *scales* package provides *ggplot2*'s internal scaling infrastructure as well as many functions for data transformation and data formatting. You can find out more about the *scales* package at this URL: `https://cran.r-project.org/web/packages/scales/scales.pdf`.

- *ggrepel*: The *ggrepel* package provides *geoms* that reposition labels to minimize label overlap. This package comes in handy when you want to add labels to your scatter plot charts. You can find more information about the *ggrepel* package at this URL: `https://cran.r-project.org/web/packages/ggthemes/ggthemes.pdf`.

Summary

In this chapter, you learned about

- A framework to use when adding R visuals to Power BI

- The *layered grammar of graphics* of which *ggplot2* is based on

- How to use geoms to add layers to your visual

- How to modify chart aesthetics using scale functions

- How to modify non-data elements of your chart using complete themes and the *theme()* function

The goal of this chapter was to introduce you to some important concepts about *ggplot2* to make it easier to understand how to build the more advanced graphics you will learn in the next chapter. Now that you know the basics of how to create R visuals in Power BI using *ggplot2*, let's take it up a notch and develop R visuals that will help you tell some amazing data stories in Power BI!

CHAPTER 2

Creating R Custom Visuals in Power BI Using ggplot2

In the previous chapter, you were introduced to the *ggplot2* package, and you were given a basic template to use for creating visuals in Power BI. In this chapter, you will be given some recipes that illustrate how expressive you can be when you leverage *ggplot2* to create R custom visuals in Power BI. This chapter contains recipes to create the following charts:

- A *callout chart* with dynamic titles and annotations

- A *bubble chart* with dynamic titles that displays five dimensions of data

- A detailed *forecast visualization* that forecasts the Wikipedia page views for Lamar Jackson and Deshaun Watson

- A *line chart* that displays GDP by Year with a *background shade* based on the political party in power for the point in time

- A *map* visualization that displays the state selected and colors the counties in that state based on the population quintile the county is in

- A *quad chart* that compares LA Lakers basketball players based on total points scored and rebounds made in the 2008–2009 season

- A *scatter plot* of the average weight by height of US women that is overlaid with a *regression line*

39

© Ryan Wade 2020
R. Wade, *Advanced Analytics in Power BI with R and Python*, https://doi.org/10.1007/978-1-4842-5829-3_2

You can make minor modifications to these recipes to use with your data. You can also be creative and take a little bit from each of the visuals to create a custom visualization that is specific to your needs. The *ggplot2* package is so rich that it will be hard to limit your creativity. Let's start by creating a *callout chart*.

Callout chart

The *horizontal bar chart* is a very common visualization. It is great for ranking categorical data based on a metric. One of the default visualizations in Power BI is a bar chart, but it is pretty basic in nature. You can do some basic enhancements like emphasize a bar by highlighting it, but you can't take things further by adding features like *dynamic subtitles*, *captions*, and *dynamic annotations* without using a clunky hack. Luckily, that is not a limitation with *ggplot2*!

In this section, you will create the *horizontal bar chart* listed in Figure 2-1.

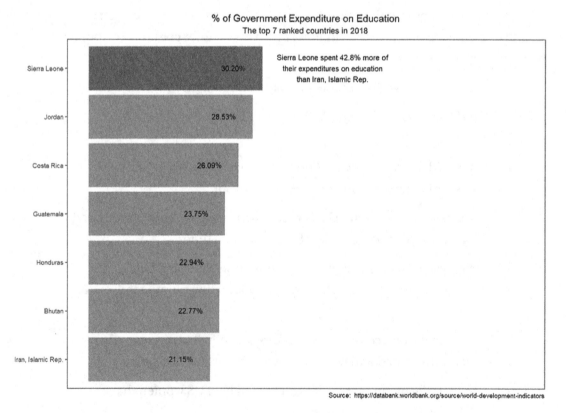

Figure 2-1. *A horizontal bar chart with a callout bar and annotation*

The preceding chart ranks the top 7 countries based on percentage of government expenditures that goes to education. It includes a *dynamic annotation* in the chart that compares the top ranked country to the seventh ranked country. Lastly, it includes a caption that tells you where the data came from. Let's go over the code needed to develop this visualization.

Step 1: Acquire the necessary data

The data set that is used for this example comes from the *World Bank Open Data*. The data set represents the percentage of government expenditures on education for each country in the world. Perform the following steps to acquire the data set:

1. Go to data.worldbank.org.

2. Go to the search bar located in the upper middle portion of the page and type *"Government expenditure on education, total (% of government expenditure)"*. After you press Enter, a list of possible options should appear and one should be exactly what you just typed. Select it and you will be taken to a page dedicated to this topic.

3. On the right side of the page, you should see a download section with options to download as *csv*, *xml*, or *Excel*. Click the *csv* option.

4. The action of the previous step will download a zip file to your computer. As of the writing, the zip file contains three files: a data file, a metadata file with country information, and a metadata file with information about the indicator. The file needed for this exercise is the data file. At the time of the writing, the file was named *API_SE.XPD.TOTL.GB.ZS_DS2_en_csv_v2_715566.csv*.

Please note that the government agencies that produce these data sets can change the format of these data sets without warning so the format may change in the future.

---

5. Now you need to import the data set into Power BI and put it in the shape it needs to be in for the visual. The *GetData* feature in Power BI will be used to perform that task. To load the contents into Power BI, go to the *Home* tab, then click *GetData* ➤ *Text/CSV*, then migrate to the location where the data file is located. Recall that the data file that was used for this example is named *API_SE.XPD.TOTL.GB.ZS_DS2_en_csv_v2_715566.csv* and it is the biggest of the three files in the zip file. The file is also located in the repo for this chapter.

6. Next, you need to make the necessary transformation to the data using *GetData (Power Query)* to get it in the shape that it needs to be in for the visualizations. The final output needs to contain the following columns with the associated data types:

Column Name	Data Type
Country	Text
Country Code	Text
Year	Whole number
Total Expend %	Decimal number

You will have a *pbix* file that contains the Power Query steps needed to get the data into the preceding shape in the code repository for this chapter. All you need to do is change the source file. Please note the steps may need to change in the future if the government entity that produced the data set decides to change the format.

Step 2: Create a slicer based on the year in the Filter pane

Drag the slicer icon to the upper right corner of the report canvas. The slicer icon is pictured in Figure 2-2.

Figure 2-2. *Image for slicer visual*

After you finish positioning the slicer, drag the *Year* field from the *ExpenditureOnEducationByCountryByYear* data set to the *Field* pane as illustrated in Figure 2-3.

Figure 2-3. *Image for the Fields pane*

Step 3: Configure the R visual in Power BI

Go to the visualization pane and drag an R visual to the report canvas. Resize the visual to the desired size. Drop the *Country, Total Expend %,* and *Year* fields from the *EducationExpenditures* table to the *Fields* pane. If the *Fields* pane is not showing, select the R visual on the report canvas and it should appear.

Step 4: Export data to R Studio for development

Select the R visual to expose the *R script editor*. You will see it at the bottom of the R visual. If it is not fully exposed, then click the upward arrow that is located on the far right corner of the *R script editor*. Once expanded, you should see the following code (Figure 2-4).

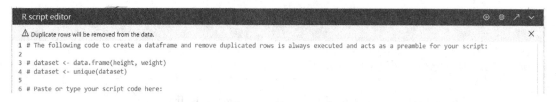

Figure 2-4. *The Power BI R script editor with the default comments*

The comments are letting you know that R took the data set you created in Power BI via the *Fields* pane and assigned it to an R object called a *data frame* that is named *dataset*. A *data frame* is a two-dimensional data set with special properties. The *data frame* is created in the preceding code in line 3. Power BI only allows data sets with unique rows to be passed to R, and it ensures that the *dataset* data frame has unique rows by using the *unique()* function in line 4.

The preceding editor is not an ideal place to do your development. Luckily, Microsoft makes it easy for you to do the development in a more desirable environment like *R Studio*. To transfer development to *R Studio*, click the 45° arrow. The arrow in question is highlighted in Figure 2-5.

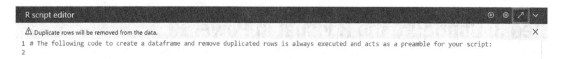

Figure 2-5. *Clicking the highlighted arrow launches your preferred IDE for a more pleasant development environment*

Clicking the arrow will open *R Studio* and will save the data set that was passed to the R visual to a temporary location on your computer. It will also generate code that will create the dataset variable that represents the data frame that was passed to the R visual in Power BI. Lastly, any R code that is in the script will be passed to *R Studio* as well. Listing 1-1 shows what will be passed to R Studio.

Listing 1-1. Code passed to R Studio from Power BI

```
# Input load. Please do not change #
`dataset` =
  read.csv('<path to input file>',
          check.names = FALSE,
          encoding = "UTF-8",
```

```
        blank.lines.skip = FALSE
    );
# Original Script. Please update your script content here
# and once completed copy below section back to the original
# editing window. The following code to create a dataframe and
# remove duplicated rows is always executed and acts as a
# preamble for your script:

# dataset <-
#    data.frame(
#      Country Name,
#      `Percent Education Expenditure Ranking`
#    )
# dataset <- unique(dataset)

# Paste or type your script code here:
```

The code. was reformatted so that it would fit nicely on the page. Lines 2–7 will display on one line which is used to read the data set that was saved to disk into an R data frame. The name *dataset* will be given to the data frame. Note that this is the same name that is used in Power BI. That was done on purpose. Using the same name enables you to use the code you will develop in *R Studio* in Power BI without having to do any refactoring.

Step 5: Load the required packages

The three packages. used in this example are *tidyverse, scales*, and *ggthemes*. The *dplyr* package from *tidyverse* is used to perform some data wrangling, *ggplot2* from *tidyverse* is used to create the visualization, *forcats* from *tidyverse* will be used to define some sorting rules, some functions from the *scales* package are used for number formatting, and the *ggthemes* package provides the *theme_few()* theme. Here is the code used to load the packages:

```
library(tidyverse)
library(scales)
library(ggthemes)
```

Step 6: Create the variables needed for the data validation test

The following code is used to create the data validation test:

```
currentColumns <- sort(colnames(dataset))
requiredColumns <- c("Country", "Total Expend %", "Year")
columnTest <- isTRUE(all.equal(currentColumns, requiredColumns))
reportYear <- unique(dataset$Year)
```

You need to ensure that the data passed to R meets the requirements for the visualization. Two tests are performed earlier to see if the requirements are met. First, a test is done to make sure the expected columns were passed to R. Second, a test is done to make sure the data set only represents one year.

The column names-based test is done in the first three lines of code. The first line creates a character vector named *currentColumns* with the names of the columns in the *dataset* data frame sorted in alphabetical order. It gets the names of the columns from *dataset* data frame using the *colnames()* function. The *sort()* function is used to sort the returned column names in alphabetical order. It needs to be in alphabetical order for comparison reasons. The next line of code creates a character vector named *requiredColumns* that contains the column names that the visual is expecting. The column names must be supplied in alphabetical order as well.

In order to test them for equality, they must not only contain the same elements, but the elements must be in the same order. To test for equality, the *all.equal()* function from base R is used. It returns *TRUE* if the two character vectors are equal; otherwise, it returns information about why they are unequal. You need a *Boolean* value to be returned. You can ensure that one is returned by wrapping the *all.equal()* function with the *isTRUE()* function. The last line in the preceding code gets the unique years contained in the *dataset* data frame. That information will be used in the next step.

Step 7: Create the data validation test

Here is the shell of the validation test:

```
if (length(reportYear) == 1 & columnTest) {

  <code to produce visual>
```

```
} else {

  plot.new()
  title("The data supplied did not meet the requirements of the
        chart.")

}
```

If the *dataset* data frame only contains one year and the *columnTest* evaluates to *TRUE,* then the code that creates the visualization is executed. Otherwise, a blank chart with the message "The data supplied did not meet the requirements of the chart." is shown.

Step 8: Add additional columns to your data set that is required for the custom R visual

Now you need to start developing the script that creates the visualization when the requirements are met. You need to enhance the *dataset* data frame with additional columns in order to create the desired visualization. The code that will be used to enhance the *dataset* data frame is shown as follows:

```
plotdata <-
  dataset %>%
  mutate(
    rank = dense_rank(desc(`Total Expend %`)),
    callout = ifelse(rank == 1, TRUE, FALSE)
    Country = fct_reorder(Country, rank, .desc = TRUE),
  ) %>%
  filter(rank <= 7)
```

This code will go where you see the `<code to produce visual>` in the template shown in Step 7. The preceding code uses the *dplyr* package from *tidyverse* to add the needed columns. The first line of code assigns the result of the succeeding code to a variable named *plotdata*. The second line of code says that you are starting with the *dataset* data frame and you are passing it as the first argument of the *mutate()* function by using the pipe *(%>%)* operator.

The pipe, *%>%*, operator comes from the *magrittr* package which is preloaded with *dplyr*. It passes the result of an expression or a value to the succeeding function. It makes code that involves chain operations much more readable.

The *mutate()* function is used to project two new columns, *rank* and *callout*, and also to change the default sort order of the *Country* column. The rank column uses the *dense_rank() window* function from *dplyr* to rank each row based on the *Total Expend % * in descending order.

The *callout* column is used to identify the country that is ranked number one. It uses the *ifelse()* function to return *TRUE* if the rank column is equal to one and return *FALSE* otherwise. The *Country* column uses the *fct_reorder()* function from *forcats* to base its sort order on the *rank* column. It can do so because the data type of the *Country* column is *factor*.

A *factor* is a special data type in R that is used for categorical data. It has many great attributes, and one of them is the ability to sort a factor column based on another column.

The *Country* field will be sorted based on its rank. The sort order defined will be used in the visualization.

Lastly, the *filter()* function is used to filter the *dataset* data frame. A filter is applied to the *rank* column to limit the *dataset* data frame to only include countries that ranks in the top 7.

Step 9: Create the variables that will be used for the dynamic portions of the chart

One of the features that is available in R but not in Power BI is the ability to dynamically add annotations to your chart. You can not only dynamically create the text you want to display on the chart but you can also dynamically control the position you want to place the text. In this step, you are creating the variables that will be used to create the dynamic annotation, and you are also defining the position of the dynamic annotation. You will also create the variables that will be used for the main chart title, the chart subtitle, and the chart caption as well. Here is the code in question:

```r
countryToAnnotate <- plotdata$`Country`[plotdata$rank == 1]
minExpenditure <- min(plotdata$`Total Expend %`)
maxExpenditure <- max(plotdata$`Total Expend %`)

minCountry <-
  plotdata$`Country`[
    which(plotdata$`Total Expend %` == minExpenditure)
  ]
minCountry <- paste(minCountry, collapse = " & ")

maxCountry <- plotdata$`Country`[
  which(plotdata$`Total Expend %` == maxExpenditure)
  ]
maxCountry <- paste(maxCountry, collapse = " & ")

mainTitle <- "% of Government Expenditure on Education"

subTitle <-
  paste(
    "The top 7 ranked countries in",
    reportYear,
    sep = " "
  )
caption <- "Source:  https://databank.worldbank.org/source/world-
development-indicators"

  label_val <-
    str_wrap(
      paste(
        maxCountry,
        "spent",
        percent((maxExpenditure/minExpenditure-1)),
        "more of their expenditures on education than",
        minCountry,
        sep = " "
      ),
      width = 35
    )
```

Here are the explanations for each variable:

- *countryToAnnotate*: This is the country that ranks number one. It will be the country that you will annotate.

- *minExpenditure*: This variable holds the result of using the *min()* function to get the min value of the *Total Expend %* column. This information will be later used to identify the country that has the minimum expenditures.

- *maxExpenditure*: This variable holds the result of using the *max()* function to get the max value of the *Total Expend %* column. This information will be later used to identify the country that has the maximum expenditures.

- *minCountry*: You identified the minimum percent expenditure in the top 7 countries in the *minExpenditure* variable. Now you need to find out which countries are associated with it. You find out that information by subsetting the *Country* column in the *plotdata* data frame to only include the countries where the *Total Expend %* is equal to the *minExpenditure*. You use the *which()* function to help you with this task. The *which()* function returns the *TRUE* indexes which can be used to tell R the indexes of the *Country* column to return. The *plotdata* data frame may contain ties which would need to be handled. That is done in the next line. Tied countries are collapsed into a scalar character vector using the *paste()* function. The elements in the newly collapse scalar character vector are separated with the & sign. You can see an example of ties if you select calendar year 2005 in the Power BI report.

- *maxCountry*: The *which()* function is used to help us get the max country name as well. Here it is being used to get the index where the *rank* column is equal to 1. There are no ties for the number one spot in this data set so you really do not need to handle them.

- *mainTitle*: This variable holds the value of a static string that will be used for the chart title.

- *subtitle*: This variable holds the value of a dynamically created string that will be used for the chart's subtitle based on the year that was selected by the report user.

- *caption*: This variable stores the website URL where the source data came from.

- *label_val*: This variable contains the result of a dynamically created chart annotation. The annotation compares the percent expenditures of the country that spent the biggest percentage of their government expenditures on education to the country that was ranked seventh. It dynamically builds the annotation using the *paste()* function by combining the *maxCountry* and *minCountry* variables along with the result of the percent calculation.

The annotation is relatively long and will not fit on the visual if you display it as one line. That problem is resolved by using the *str_wrap()* function from the *stringr* package. This function provides a wealth of formatting options, one of which is the ability to control the maximum width of a line. You set this argument to 35, in this example, which ensures that the annotation will fit on the chart.

Step 10: Start building the chart by defining the ggplot( ) function

The *ggplot()* function is defined here:

```
p <-
  ggplot(
    data = plotdata,
    aes(x = `Country`, y = `Total Expend %`, fill = callout,
        label = percent(`Total Expend %`)
    )
  )
```

The purpose of the *x* and *y* aesthetic is self-explanatory. The *fill* aesthetic tells *ggplot()* what column (variable) you want to base the coloring of the bars on, and the *label* aesthetic tells R what column will be used to add labels to the chart.

Step 11: Add a column chart layer to the R visual

In this step, you add the first layer to the chart which is a column chart using the *geom_col()* geom. You have the option to use this or the *gom_bar()* geom. You chose *geom_col()* because the default for the stat argument is *stat_identity* so the true value of the y aesthetic will be displayed and not a count of the number of elements in *y*. If we used *geom_bar()*, you would have to explicitly tell *ggplot()* that you want the stat argument set to *stat_identity*:

```
p <-
  ggplot(
    data = plotdata,
    aes(x = `Country`, y = `Total Expend %`, fill = callout,
        label = percent(`Total Expend %`)
    )
  ) +
  geom_col()
```

Step 12: Add a text layer to the R visuals

The following code takes what was built up to Step 11 and adds a layer to the chart for the labels:

```
p <-
  ggplot(
    data = plotdata,
    aes(x = `Country`, y = `Total Expend %`, fill = callout,
        label = percent(`Total Expend %`)
    )
  ) +
  geom_bar(stat="identity", aes(fill = callout)) +
  geom_text(nudge_y = -0.05)
```

In this step, you add another layer to the chart. You add the labels to the chart using the *geom_text()* geom. The required arguments for the *geom_text()* geom are the *x, y,* and *label* arguments. All three are inherited from the aesthetics defined in the *ggplot()* function. The *x* and *y* arguments are used to define position of the labels, and the *label* argument is used to define what will be displayed. In this example, you are displaying the *Total Expend %* column formatted as a percent. The natural position of the labels is partially outside the end of the columns of the chart. You need them to be contained totally in the column chart so that the labels won't overlap with the annotation that will be added later. You move the labels from partially outside the columns to totally inside the columns via the *nudge_y* argument. Figure 2-6 shows how the labels display without using the *nudge_y* argument.

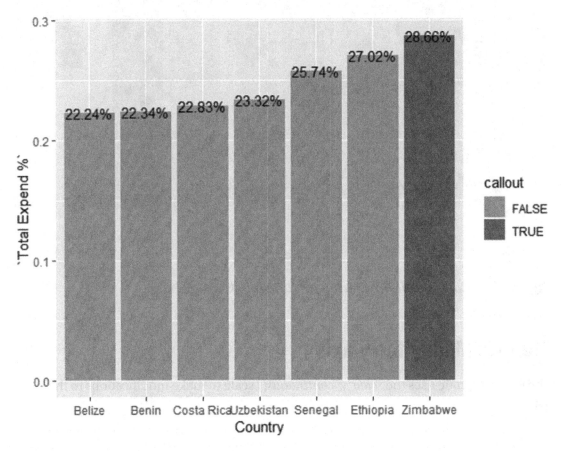

Figure 2-6. *Callout column chart with data labels*

Figure 2-7 shows what the labels look like when you set the *nudge_y* equal to –0.05.

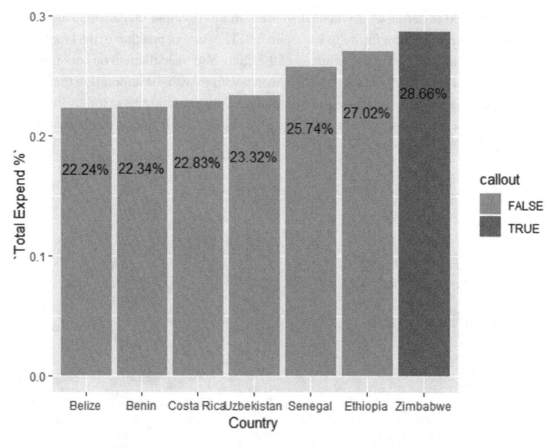

Figure 2-7. *Callout column chart with data labels inside bars*

Step 13: Modify the y axis

The following code uses the *scale_y_continuous()* scale to make modifications to the *y* axis:

```
p <-
  ggplot(
    data = plotdata,
    aes(x = `Country`, y = `Total Expend %`, fill = callout,
        label = percent(`Total Expend %`)
    )
  ) +
```

```
geom_bar(stat="identity", aes(fill = callout)) +
geom_text(nudge_y = -0.05) +
scale_y_continuous(
    limits = c(0,0.75),
    labels = NULL,
    breaks = NULL
)
```

You explicitly set the limits of the *y* scale to the range 0 to 0.75. You do this to give the report user a consistent experience. By default, like many visualization tools, *ggplot2* will adjust the scales based on the data. This can visually misrepresent important differences in the data between years. So, you prevent rescaling by explicitly defining the range. The *labels* and *breaks* are used to help report users interpolate values of the columns in the chart. Since you are providing those values explicitly, you don't need labels and breaks. So, they are removed by setting the labels and breaks argument to *NULL*.

Step 14: Convert chart from a vertical column chart to horizontal bar chart

This action is done by adding the *coord_flip()* function to the next line of the script as illustrated here:

```
p <-
  ggplot(
    data = plotdata,
    aes(x = `Country`, y = `Total Expend %`, fill = callout,
        label = percent(`Total Expend %`)
    )
  ) +
  geom_bar(stat="identity", aes(fill = callout)) +
  geom_text(nudge_y = -0.05) +
  scale_y_continuous(
      limits = c(0,0.75),
      labels = NULL,
      breaks = NULL
  ) +
  coord_flip()
```

The *coord_flip()* function causes the *x* axis to become the *y* axis and the *y* axis to become the *x* axis. The result is a horizontal bar chart illustrated in Figure 2-8.

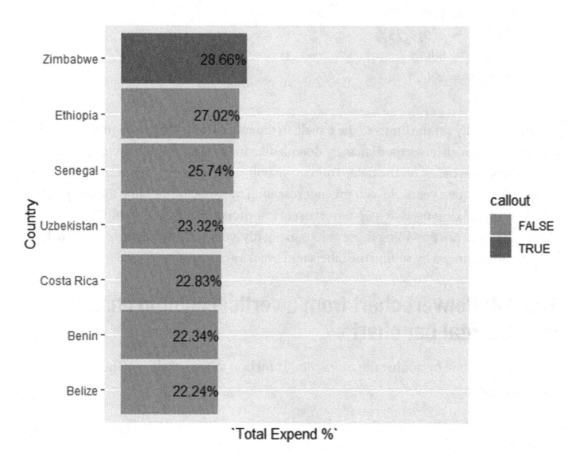

Figure 2-8. *Previous column chart switched to horizontal bar chart*

Step 15: Add a dynamic annotation to the R visuals

You add an annotation layer to the chart using the *annotate()* function as illustrated here:

```
p <-
  ggplot(
    data = plotdata,
    aes(x = `Country`, y = `Total Expend %`, fill = callout,
        label = percent(`Total Expend %`)
    )
  ) +
```

```
geom_bar(stat="identity", aes(fill = callout)) +
geom_text(nudge_y = -0.05) +
scale_y_continuous(
    limits = c(0,0.75),
    labels = NULL,
    breaks = NULL
) +
coord_flip() +
annotate(
    "text",
    label = label_val,
    x = countryToAnnotate[1],
    y = maxExpenditure + 0.12
)
```

You want to add a text annotation to the chart so the first argument is set to *"text"*. Next, you set the *label_val* variable to the *label* argument because that variable contains the text for the annotation. Lastly, you define the position of the annotation using the *x* and *y* arguments. You set *x* to the name of the country you want to annotate. Setting *x* to the country name aligns the annotation at the position of the *x axis* where the specified country is. You set the position on the *y* axis to a position that is a few units higher than the expenditure value of the country so that the annotation will not overlay the country's column. Note that because you used the *coord_flip()* function earlier, it appears you are working on the *x axis* when you are actually working on the *y axis* and vice versa. The result of adding the annotation layer is illustrated in Figure 2-9.

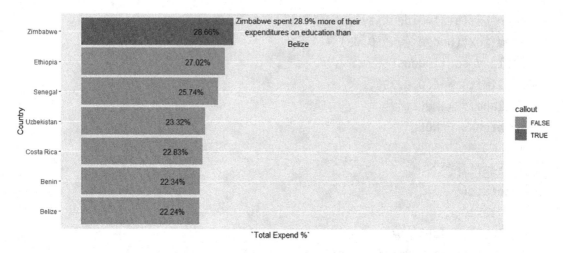

Figure 2-9. *Horizontal bar chart with annotation*

Step 16: Add the dynamic titles and caption to the R visual

Next, you add the *chart title, subtitle,* and *caption* to the chart using the *labs()* function along with the *mainTitle, subTitle,* and *caption* variable as illustrated in the following code:

```
p <-
  ggplot(
    data = plotdata,
    aes(x = `Country`, y = `Total Expend %`, fill = callout,
        label = percent(`Total Expend %`)
    )
  ) +
  geom_bar(stat="identity", aes(fill = callout)) +
  geom_text(nudge_y = -0.05) +
  scale_y_continuous(
    limits = c(0,0.75),
    labels = NULL,
    breaks = NULL
  ) +
```

```
coord_flip() +
annotate(
    "text",
    label = label_val,
    x = countryToAnnotate[1],
    y = maxExpenditure + 0.12
) +
labs(
    title = mainTitle,
    subtitle = subTitle,
    caption = caption
)
```

Setting the labs function is self-explanatory so no explanation is needed.

Step 17: Remove labels from x axis and y axis

You don't need to explicitly define the *x* and *y* axis because of the labels you added to the chart. You remove the *x* and *y* labels by setting the *label* argument in the *xlab()* and *ylab()* functions to *NULL* as illustrated in the last line in the following code:

```
p <-
  ggplot(
    data = plotdata,
    aes(x = `Country`, y = `Total Expend %`, fill = callout,
        label = percent(`Total Expend %`)
    )
  ) +
  geom_bar(stat="identity", aes(fill = callout)) +
  geom_text(nudge_y = -0.05) +
  scale_y_continuous(
    limits = c(0,0.75),
    labels = NULL,
    breaks = NULL
  ) +
```

```
coord_flip() +
annotate(
    "text",
    label = label_val,
    x = countryToAnnotate[1],
    y = maxExpenditure + 0.12
) +
labs(
    title = mainTitle,
    subtitle = subTitle,
    caption = caption
) +
xlab(label = NULL) +
ylab(label = NULL)
```

Step 18: Remove legend

If you refer back to Figure 2-10, you will notice a legend for *callout*. The callout column is associated with the *fill* argument which is used to determine the color of the bars. You use it to color the country ranked number one differently than the other countries. A legend is not needed to explain the colors, so it is removed. You do so by setting the *fill* argument in the *guides()* function to *FALSE*. Doing so removes the legend from the chart because the legend is for the *fill* scale. The code used to perform the task is in the last line in the following script:

```
p <-
  ggplot(
    data = plotdata,
    aes(x = `Country`, y = `Total Expend %`, fill = callout,
        label = percent(`Total Expend %`)
    )
  ) +
  geom_bar(stat="identity", aes(fill = callout)) +
  geom_text(nudge_y = -0.05) +
```

```
scale_y_continuous(
   limits = c(0,0.75),
   labels = NULL,
   breaks = NULL
) +
coord_flip() +
annotate(
   "text",
   label = label_val,
   x = countryToAnnotate[1],
   y = maxExpenditure + 0.12
) +
labs(
   title = mainTitle,
   subtitle = subTitle,
   caption = caption
) +
xlab(label = NULL) +
ylab(label = NULL) +
guides(fill=FALSE)
```

Step 19: Change the look and feel of the visual using theme_few()

Next, you change the look of the chart using principles by the data visualization expert, Stephen Few, via the *theme_few()* theme from the *ggthemes* package as illustrated in the last line in the following code:

```
p <-
  ggplot(
    data = plotdata,
    aes(x = `Country`, y = `Total Expend %`, fill = callout,
        label = percent(`Total Expend %`)
    )
  ) +
```

```
geom_bar(stat="identity", aes(fill = callout)) +
geom_text(nudge_y = -0.05) +
scale_y_continuous(
    limits = c(0,0.75),
    labels = NULL,
    breaks = NULL
) +
coord_flip() +
annotate(
    "text",
    label = label_val,
    x = countryToAnnotate[1],
    y = maxExpenditure + 0.12
) +
labs(
    title = mainTitle,
    subtitle = subTitle,
    caption = caption
) +
xlab(label = NULL) +
ylab(label = NULL) +
guides(fill=FALSE) +
theme_few()
```

Step 20: Center align titles

The *theme()* function is used in the last line of the following code to handle this task:

```
p <-
  ggplot(
    data = plotdata,
    aes(x = `Country`, y = `Total Expend %`, fill = callout,
        label = percent(`Total Expend %`)
    )
  ) +
```

```
geom_bar(stat="identity", aes(fill = callout)) +
geom_text(nudge_y = -0.05) +
scale_y_continuous(
    limits = c(0,0.75),
    labels = NULL,
    breaks = NULL
) +
coord_flip() +
annotate(
    "text",
    label = label_val,
    x = countryToAnnotate[1],
    y = maxExpenditure + 0.12
) +
labs(
    title = mainTitle,
    subtitle = subTitle,
    caption = caption
) +
xlab(label = NULL) +
ylab(label = NULL) +
guides(fill=FALSE) +
theme_few() +
theme(
    plot.title = element_text(hjust = 0.5),
    plot.subtitle = element_text(hjust = 0.5)
)
```

The *theme()* function is used to modify individual components of the chart's theme. The chart's theme represents the non-data components of the chart. You can type ?theme() in the console to access the help file for *theme()* to see what components are available for modification and the element function that is associated with it. The *element_text()* function is the element function associated with *plot.title* and *plot.subtitle* components. Setting the *hjust* argument of *element_text()* to 0.5 centers the text.

Step 21: Add code to Power BI

The entire R script is in Listing 2-2. Copy the code and paste it into the R script editor associated with this visual in Power BI. There will be no need for refactoring if you followed the steps correctly.

Listing 2-2. The callout chart R script

```
library(tidyverse)
library(scales)
library(ggthemes)

currentColumns <- sort(colnames(dataset))
requiredColumns <- c("Country", "Total Expend %", "Year")
columnTest <- isTRUE(all.equal(currentColumns, requiredColumns))
reportYear <- unique(dataset$Year)

if (length(reportYear) == 1 & columnTest) {

  plotdata <-
    dataset %>%
    mutate(
      rank = dense_rank(desc(`Total Expend %`)),
      `Country` = fct_reorder(`Country`, rank, .desc = TRUE),
      callout = ifelse(rank == 1, TRUE, FALSE)
    ) %>%
    filter(rank <= 7)

  countryToAnnotate <- plotdata$`Country`[plotdata$rank == 1]
  minExpenditure <- min(plotdata$`Total Expend %`)
  maxExpenditure <- max(plotdata$`Total Expend %`)
  minCountry <-
    plotdata$`Country`[
      which(plotdata$`Total Expend %` == minExpenditure)]
  minCountry <- paste(minCountry, collapse = " & ")
  maxCountry <-
    plotdata$`Country`[
      which(plotdata$`Total Expend %` == maxExpenditure)]
  maxCountry <- paste(maxCountry, collapse = " & ")
```

```
mainTitle = "% of Government Expenditure on Education"
subTitle =
  paste(
    "The top 7 ranked countries in",
    reportYear,
    sep = " ")
caption =
  "Source: https://databank.worldbank.org/source/world-development-
  indicators"

label_val <-
  str_wrap(
    paste(maxCountry, "spent",
          percent((maxExpenditure/minExpenditure-1)),
          "more than", minCountry, sep = " "),
    width = 25
  )

p <-
  ggplot(
    data = plotdata,
    aes(x = `Country`, y = `Total Expend %`, fill = callout,
        label = percent(`Total Expend %`))
  ) +
  geom_bar(stat="identity", aes(fill = callout)) +
  geom_text(nudge_y = -0.05) +
  scale_y_continuous(
    limits = c(0,0.75), labels = NULL, breaks = NULL) +
  coord_flip() +
  annotate(
    "text",
    label = label_val,
    x = countryToAnnotate[1],
    y = maxExpenditure + 0.1) +
```

```
  labs(
    title = mainTitle,
    subtitle = subTitle,
    caption = caption
  ) +
  xlab(label = NULL) +
  ylab(label = NULL) +
  guides(fill=FALSE) +
  theme_few() +
  theme(
    plot.title = element_text(hjust = 0.5),
    plot.subtitle = element_text(hjust = 0.5))
  p
} else {
  plot.new()
  title("The data supplied did not meet the requirements of the chart.")
}
```

Bubble chart

The *bubble chart* is another popular chart type. You can think of them as a scatter plot
with an additional attribute that uses size to represent another dimension of data. So
with a traditional *bubble chart*, you are able to show up to five dimensions of data:

- The *x* and *y* of each bubble represent two dimensions.

- The *fill* (the color inside the bubble) represents a dimension.

- The *size* represents a dimension.

- The *labels* represents a dimension.

You can take things to the next level using *ggplot2*. You can color the border of the
bubbles which will enable you to show six dimensions of data in your *bubble chart*! In
this example, you will use data from a fictitious professional football league that plots
running backs on six dimensions of data: *name, height, weight, total rushing yards,
conference,* and *division*. The chart that you will create in this exercise is shown in
Figure 2-11. Let's go over the steps to create the chart.

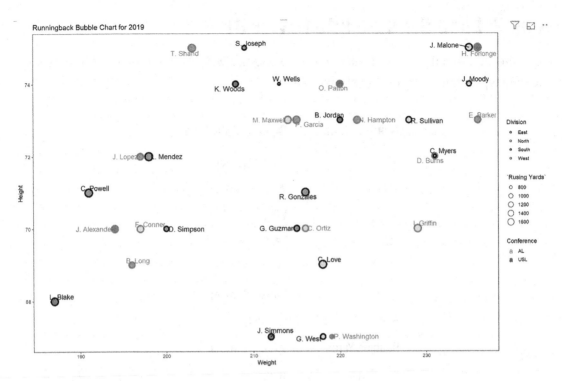

Figure 2-10. *A bubble chart that displays six dimensions of data*

Step 1: Acquire the necessary data

The data set used in this example is a fictitious data set of a pretend professional football league. The league has two conferences: *AL (American League)* and *USL (United States League)*. Each conference contains four divisions: *North, South, East*, and *West*. The data set contains the names, height, weight, rushing yards, conference, and division of the running backs in this fictitious league. The name of the file that contains the data set is *FakeFootballLeagueData.csv* and can be found in the code repository for this chapter.

Step 2: Load the data into Power BI

Load the data into the Power BI data model using *GetData*. The following table shows the data types that each column should have:

Column Name	Column Data Type
ID	Whole Number
Year	Whole Number
Key	Whole Number
FN	Text
LN	Text
Weight	Whole Number
Height	Whole Number
Rushing Yards	Whole Number
Division	Text
Conference	Text

The transformation steps needed to get the data in the preceding format are in the PBI_FakeFootballLeague.pbix file.

Step 3: Create a filter slicer based on the year

Create a report slicer based on the year field by first dragging the *Year* field to the report canvas, then selecting the *Slicer* visual in the *Visualization* pane. Make sure the *Year* field you drag is highlighted before you select the *Slicer* visual to convert it into a slicer. The slicer visual is the one that includes a picture of a filter in the image.

Step 4: Do the initial R visual configuration

Go to the visualization pane and drag an R visual to the report canvas. Resize the visual to the desired size. Drop the *ID, Year, FN, LN, Weight, Height, Rushing Yards, Division,* and *Conference* fields from the *FakeFootballLeague* table to the *Fields* pane. If the Fields pane is not showing, select the R visual and it should appear.

Step 5: Export data to R Studio for development

Click the 45° arrow that is located on the title bar of the *R script editor*. This action will export the data frame passed to R and the R starter code to *R Studio*. Please refer to Step 4 in the *"Callout chart"* section if you need a more detailed explanation.

Step 6: Load the required packages

The following code is used to load the required packages:

```
library(tidyverse)
library(ggrepel)
library(ggthemes)
```

The *dplyr* and *ggplot2* packages are used in this script from *tidyverse*. The *dplyr* package is used to wrangle the data passed to R from Power BI into the shape it needs to be in, and *ggplot2* is used to create the visual. The *ggrepel* package is used to re-position the labels of the data points in the visual format, and the *ggthemes* package is used to change the display of the non-data components of the visual.

Step 7: Create the variables needed for the data validation test

```
currentColumns <- sort(colnames(dataset))
requiredColumns <- c("Conference", "Division", "FN", "Height",
                     "ID", "LN", "Rusing Yards", "Weight",
                     "Year")
columnTest <- isTRUE(all.equal(currentColumns, requiredColumns))
reportYear <- unique(dataset$Year)
```

The *currentColumns* variable is used to hold the column names of the data frame passed to R from Power BI. The names are sorted in alphabetical order for comparison reasons. The *requiredColumns* variable is used to hold the names of the required columns. They are also listed in alphabetical order for comparison reasons. The *columnTest* variable holds the results of the comparison test. It uses the *all.equal()* function to compare the *currentColumns* variable to the *requiredColumns* variable for equality. If those two variables are equal, then *TRUE* is returned; otherwise, information about what was not equal between the two variables is returned. You need a Boolean response, so you wrap the *all.equal()* function with *isTRUE()* function which will return *TRUE* when *all.equal()* returns *TRUE* and it will return *FALSE* when *all.equal()* returns something other than *TRUE*. Next, the *unique()* function is used to get a list of unique elements from the *Year* column.

Step 8: Create the data validation test

The template used to perform the validation is as follows:

```
if(length(reportYear) == 1 & columnTest) {

    <code to create the visual>

} else{

    plot.new()
    title("The data supplied did not meet the requirements of the
          chart.")

}
```

The *if* statement performs two tests. First, it tests to see if the length of the *reportYear* vector is equal to one. Second, it tests to see if the *columnTest* variable evaluates to *TRUE*. If both of those conditions are met, then the code that creates the visual is executed. Otherwise, code is executed that creates a blank chart with a message giving the user information about why the expected visual was not returned.

Step 9: Define the colors for the conferences and conference divisions

One of the useful features of *ggplot2* is that *ggplot2* makes it easy to customize the colors you want to use in a chart. You can define charts using a color name that *ggplot2* recognizes, or you can use the *hexadecimal* representation of the color if you know it. Here, you are using the latter to define your colors via *named character vectors*. You use *named character vectors* to define the colors of the division and of the conferences. The name of the elements in the *named character vectors* represents the division or conference, and the value of the element in the *named character vectors* is the hexadecimal number that represents the color you want to use. Here is the code that you use to define the colors for the divisions and conferences:

```
divisionColors <- c("East"="#56B4E9", "West"="#33FAFF",
                     "North"="#F0E442", "South"="#8B8D8D")
conferenceColors <- c("USL"="#000000", "AL"="#FC4E07")
```

Here is the URL to a website that you can use to get the hexadecimal number of various colors: www.hexcolortool.com/#010328. This site can also be use when you want to the hexadecimal value of a color you want to use in native a Power BI visual.

Step 10: Dynamically define the chart titles

Here is the code that creates the chart title:

```
chartTitle <- paste("Runningback Quad Chart for",
                     reportYear,
                     sep = " "
         )
```

The code concatenates the string "Runningback Quad Chart for" with the report year selected by the report user to dynamically create the chart title.

Step 11: Create the chart's data set

The data set that was passed to Power BI is not totally ready for the chart. You want to add a label to the data points in the chart for each running back. The labels would be too long if you used both their first and last name, so you need a way to shorten them. A good method to use would be to take the initial of the first name and concatenate it with the last name. That is accomplished in the following code:

```
chartData <-
  dataset %>%
  mutate(Name = paste0(str_sub(FN,1,1),". ", LN))
```

The preceding code starts with the *dataset* data frame, then projects a new column to it named *Name*. It is based on the concatenation of three strings using the *paste0* function. The first string uses the *str_sub()* function from the *stringr* package to extract the first character from the *FN* column. The second string is " . " and the third string is the *LN*. So, if FN was John and if the LN was Doe, the result would be *J. Doe*.

Step 12: Start the chart by defining the ggplot function

The bubble chart that you are creating will have six dimensions. The six dimensions are

- x coordinate of the chart to represent the running back's weight

- y coordinate to represent the running back's height

- Size of the bubble to represent the running back's total rushing yards

- The fill (inside color) of the chart to represent the running back's division

- The color (border of the bubble) to represent the running back's conference

- The label to represent the running back's name

You can set these dimensions using the *aes()* function in the *ggplot()* function as illustrated here:

```
p <- ggplot(
  chartData,
  aes(
    x = Weight, y = Height, size = `Rusing Yards`,
    color = Conference, fill = Division, stroke = 2,
    label = Name
  )
)
```

Step 13: Add the layer for your bubble chart using the geom_point geom

There are 26 shapes that you can use in the *geom_point()* geom. Figure 2-11 is a visual that depicts them.

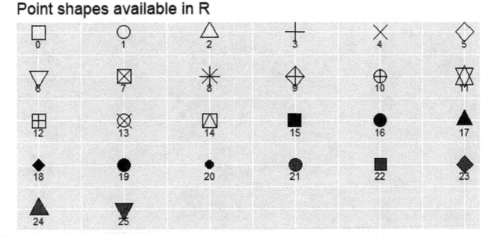

Figure 2-11. *Shows the shapes available in geom_point(). Visual was created using the ggplot package*

There are three shapes that can be used to create bubble charts. They are shapes *1,* *16,* and *21*. Only shape 21 allows you to control both the color (the border of the circle) and the fill (the inside shade) of the point. Only the color is available in shape *1,* and only the fill is available in shape *16*. The code that was used to add the *geom_point()* layer to the visual using *shape 21* is listed here:

```
p <- ggplot(
  chartData,
  aes(
    x = Weight, y = Height, size = `Rusing Yards`,
    color = Conference, fill = Division, stroke = 2,
    label = Name
  )
) +
geom_point(shape = 21)
```

The result is the visual in Figure 2-12.

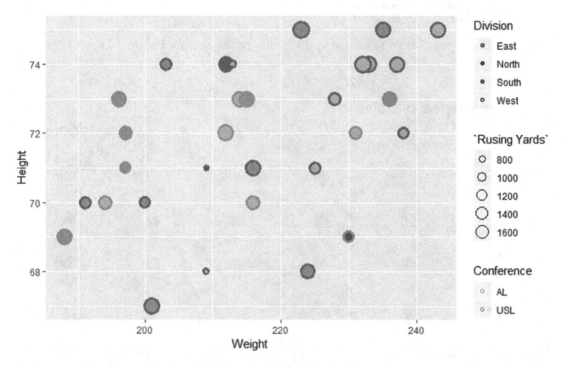

Figure 2-12. *The initial bubble chart using shape 21*

You have the basics of the bubble chart built. Note that the *geom_point()* geom inherited the aesthetics from the *ggplot()* function so you did not have to explicitly define them. Now, you need to make some cosmetic changes to make the chart visibly more appealing.

Step 14: Add labels to the bubble chart

You have bubbles scattered on the chart, but you don't know which athlete the bubbles represent. You can add text to the visual using the *geom_text_repel()* geom from the *ggrepel* package. The *geom_text_repel()* geom tries to prevent overlapping when it adds the labels. Here is the code with the *geom_text_repel()* layer added:

```
p <- ggplot(
  chartData,
  aes(
```

```
    x = Weight, y = Height, size = `Rusing Yards`,
    color = Conference, fill = Division, stroke = 2,
    label = Name
  )
) +
  geom_point(shape = 21) +
  geom_text_repel(size = 5)
```

The result of adding the *geom_text_repel()* layer is listed in Figure 2-13.

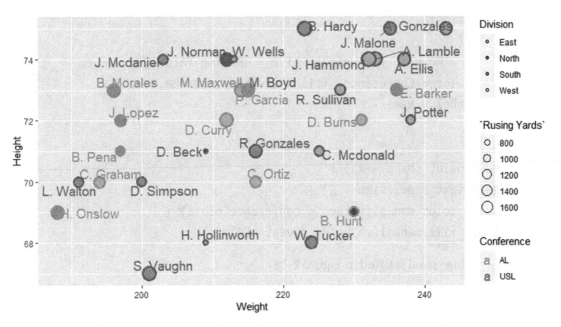

Figure 2-13. *The bubble chart with added labels*

Note that we were able to control the text size of the label via the *size* argument. Without the size argument added, the font of the labels would vary based on the size of the bubble.

Step 15: Change the color of the bubble's border and change the color of the bubble's fill

Next, you change the color of the bubbles using the predefined colors in the *conferenceColors* named character vector. You do so by setting *conferenceColors* to the *values* argument in the *scale_color_manual()* scale. You also need to set the *fill* with the predefined colors in the *divisionColors* variable. You do so by setting *divisionColors* to the *values* argument in the *scale_fill_manual()* scale. The code with the two new scales is listed here:

```
p <- ggplot(
      chartData,
      aes(
        x = Weight, y = Height, size = `Rusing Yards`,
        color = Conference, fill = Division, stroke = 2,
        label = Name
      )
    ) +
    geom_point(shape = 21) +
    geom_text_repel(size = 5) +
    scale_color_manual(values = conferenceColors) +
    scale_fill_manual(values = divisionColors)
```

The resulting visual is listed in Figure 2-14.

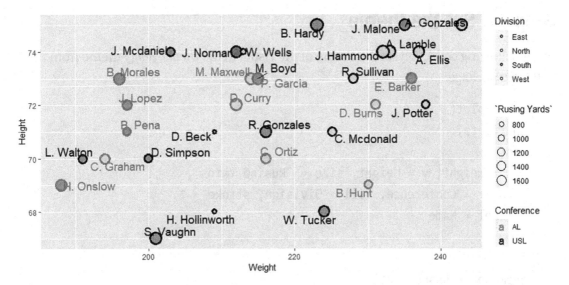

Figure 2-14. *Bubble chart with custom colors*

Step 16: Create the ggtitle

You pass the *chartTitle* variable to the *ggtitle()* function to add the title to the chart as shown in the last line in the following code:

```
p <- ggplot(
  chartData,
  aes(
    x = Weight, y = Height, size = `Rusing Yards`,
    color = Conference, fill = Division, stroke = 2,
    label = Name
  )
) +
  geom_point(shape = 21) +
  geom_text_repel(size = 5) +
  scale_color_manual(values = conferenceColors) +
  scale_fill_manual(values = divisionColors) +
  ggtitle(chartTitle)
```

Step 17: Set the theme

In the following code, you set the theme of the visual to the *theme_few()* theme from the *ggthemes* package:

```
p <- ggplot(
  chartData,
  aes(
    x = Weight, y = Height, size = `Rusing Yards`,
    color = Conference, fill = Division, stroke = 2,
    label = Name
  )
) +
  geom_point(shape = 21) +
  geom_text_repel(size = 5) +
  scale_color_manual(values = conferenceColors) +
  scale_fill_manual(values = divisionColors) +
  ggtitle(chartTitle) +
  theme_few()
```

Step 18: Add code to Power BI

You now have a functional script that produces the desired visual. The full script is as follows. Add the script in Listing 2-3 to the *R script editor* associated with the visual in Power BI. The visual will be responsive to the *Year* filter used in the report.

Listing 2-3. The bubble chart R script

```
library(tidyverse)
library(ggrepel)
library(ggthemes)

currentColumns <- sort(colnames(dataset))
requiredColumns <-
    c("Conference", "Division", "FN", "Height", "ID", "LN", "Rusing Yards",
    "Weight", "Year")
```

```r
columnTest <- isTRUE(all.equal(currentColumns, requiredColumns))
reportYear <- unique(dataset$Year)

if(length(reportYear) == 1 & columnTest) {

  chartData <-
    dataset %>%
    mutate(Name = paste0(substring(FN,1,1),". ", LN))

  divisionColors <-
      c("East"="#56B4E9", "West"="#009E73", "North"="#F0E442",
        "South"="#0072B2")
  conferenceColors <- c("USL"="#000000", "AL"="#FC4E07")

  chartTitle <-
      paste("Runningback Quad Chart for", reportYear, sep = " ")

  p <- ggplot(
          chartData,
          aes(
            x = Weight, y = Height, size = `Rusing Yards`,
            color = Conference, fill = Division, stroke = 2,
            label = Name
          )
        ) +
        geom_point(shape = 21) +
        geom_text_repel(size = 5) +
        scale_color_manual(values = conferenceColors) +
        scale_fill_manual(values = divisionColors) +
        ggtitle(chartTitle) +
        theme_few()

  p

} else{

  plot.new()
  title("The data supplied did not meet the requirements of the
        chart.")
}
```

Forecast

Forecasting business outcomes has been a technique used by companies for many years. Many business intelligence tools have forecast visualizations you can add to your dashboards that will display a forecast, but they are very limited in the amount of information they give. They are limited because they typically only show the forecast, but they leave out important components that influence a forecast. Those components include *trend* and *seasonality*. The forecast *trend* is used to show if the data is decreasing or increasing over a long period of time. The forecast *seasonality* is used to show variations in the data caused by the time of occurrence. Fortunately for us, the data science team at Facebook developed a package for R called *Prophet* that not only enables you to create robust forecast visual but also gives you the ability to create visuals that give you valuable information about the different components that make up the forecast.

In this exercise, you will build a visual that was inspired by the example used in the *Prophet* documentation. The example in *Prophet* created a visualization that forecasted Wikipedia page views for Peyton Manning. Peyton is a former NFL quarterback. You will build on what was done in the *Prophet* example by building an interactive visualization that will enable you to switch between two current NFL quarterbacks, Lamar Jackson and Deshaun Watson. The visualization will not only have a chart for the forecast but it will also have a chart to show the trend of the forecast and charts to show weekly and yearly seasonality. An image of the visual is illustrated in Figure 2-15.

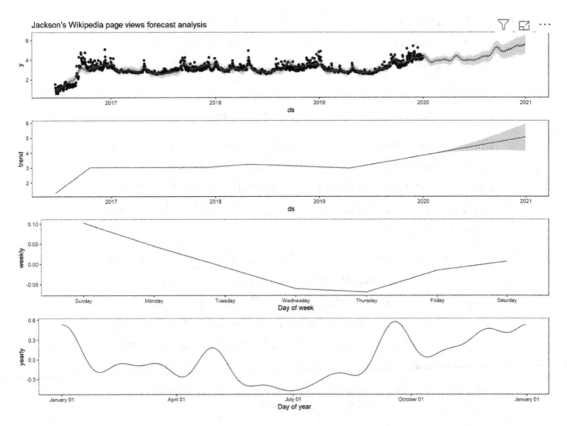

Figure 2-15. *Forecast visual produced using Prophet and ggplot2*

Let's go over the steps needed to create this custom R visual!

Step 1: Acquire the necessary data

You need to acquire the data that contains the daily Wikipedia views for Lamar Jackson and Deshaun Watson. Here are the steps you need perform to acquire the data. Note that you will need to go through these steps twice, once for Lamar Jackson and once for Deshaun Watson:

1. Go to `https://en.wikipedia.org/wiki/Wikipedia:Web_statistics_tool` to access the *Wikipedia Web Statistics Tool*.

2. Click toollabs:pageviews link in the *Current tools* section.

3. Click *X Clear* above the search bar to clear the page names and put the name of the page you want to test. On the first iteration, use *Lamar Jackson,* and on the second iteration, use *Deshaun Watson.*

4. Click the *Dates* textbox, then go to custom range.

5. Choose a data range that includes a minimum of a 2-year period. That requirement is needed so that the *Prophet* package will have enough information to detect yearly seasonality trends. In this example, I chose 1/1/2018 for my beginning date and 12/31/2019 for my ending date.

6. Click the downward arrow in the *Download* button located below the *Pages* textbox and select the *CSV* option. That action will initiate the download process, and it will save the results of your query in a *csv* format on your computer. The default name does not give you information about the subject of your query. Rename the file with a name that is representative of the contents of the file. I prefixed each file with the initials of the subject. I used *lj* for Lamar Jackson and *dw* for Deshaun Watson.

7. Next, you need to load the data set for Lamar Jackson and Deshaun Watson into Power BI, make some data transformations required for the visualization, then combine them into one data set. Power Query will be used to do that. The final data set needs to contain the following fields with the corresponding data types:

Column Name	Data Type
Date	Text
Page Views	Whole number
Quarterback	Text
Page Views (Log10)	Decimal number

The transformations performed using Power Query to get the data in the preceding format can be found in the *pbix* file located in the code repo for this chapter.

Step 2: Create a slicer based on quarterback

Drag the *Quarterback* field to the upper right portion of the report canvas. Next, go to the *Visualization* pane in the upper right corner and change the visualization to a *Slicer* visual. The *Slicer* visual comes pre-packaged in Power BI, and it is the one that includes a picture of a filter in the image.

Step 3: Configure the R visual

Go to the visualization pane and drag an R visual to the report canvas. Resize the visual to the desired size. Drop the *Date, Page Views, Page Views (Log10),* and *Quarterback* fields from the *WikipediaPageViews* table into the *Fields* pane. If the Fields pane is not showing, select the *R visual* and it should appear.

Step 4: Export data to R Studio for development

Click the 45° arrow that is located on the title bar of the *R script editor*. This action will export the data frame passed to the *R visual* and the R starter code to *R Studio*. Please refer to Step 4 in the *"Callout chart"* section if you need a more detailed explanation.

Step 5: Load the required packages to the script

Here is the code needed to load the required packages for the script:

```
library(tidyverse)
library(prophet)
library(ggthemes)
library(gridExtra)
library(lubridate)
```

The *dplyr* package from *tidyverse* is used to perform some data wrangling tasks, the *ggplot2* package from *tidyverse* is used to create the visualization, the *prophet* package is used to perform the forecast and help with creating the visualization, the *gridExtra* package is used to enable multiple chart display in a single visual, and the *lubridate* package is used to perform some date manipulation tasks.

Step 6: Create the variables needed for the data validation test

The code used to create the variables for the validation is listed here:

```
currentColumns <- sort(colnames(dataset))
requiredColumns <- c("Date", "Page Views", "Page Views (Log10)",
                     "Quarterback")
columnTest <- isTRUE(all.equal(currentColumns, requiredColumns))

qb <- unique(dataset$Quarterback)
```

The *currentColumns* character vector variable is used to hold the column names of the data frame passed to R from Power BI. The names are sorted in alphabetical order for comparison reasons. The *requiredColumns* character vector variable is used to hold the names of the required columns. They are also listed in alphabetical order for comparison reasons. The *columnTest* variable holds the results of the comparison test. It uses the *all.equal()* function to compare the *currentColumns* variable to the *requiredColumns* variable for equality. If those two variables are equal, then *TRUE* is returned; otherwise, information about what was not equal between the two variables is returned. You need a Boolean response so you wrap the *all.equal()* function with the *isTRUE()* function which will return *TRUE* when *all.equal()* returns TRUE and it will return *FALSE* when *all.equal()* returns something other than TRUE. The last thing that is done is you get all the unique elements in the *Quarterback* column and assign the result to the *qb* variable.

Step 7: Create the data validation test

The template used to execute the validation test is as follows:

```
if (length(qb) == 1 & columnTest) {

    <code to produce visual>

} else {

  plot.new()
  title("The data supplied did not meet the requirements of the
        chart.")

}
```

The visual needs a data set that only contains one quarterback and has the required columns. The first test checks to see if the length of the *qb* variable is 1. If it is equal to 1, then it means that there is only one quarterback in the data set so that test is passed. The second test checks to see if the data frame passed from Power BI contains the required columns. If the *columnTest* variable is equal to *TRUE*, then that test is passed. If both tests evaluate to *TRUE*, then R will attempt to execute the code that creates the visual. Otherwise, code is executed that creates a blank chart with a message containing information about why the expected visual was not created.

Step 8: Create the dynamic chart title

The following code creates a variable that will hold a dynamically created chart title based on the quarterback's name:

```
chartTitle <-
    paste0(qb, "'s Wikipedia page views forecast analysis")
```

It concatenates the Quarterback's name with the string, " 's Wikipedia page views forecast analysis", without any delimiters. The *paste0()* function is similar to the *paste()* function. It is used when you want to concatenate strings without using a delimiter.

Step 9: Create the data set that is needed to generate the forecast

```
dfPageViews <-
  dataset %>%
  transmute(
      Quarterback,
      ds = ymd(Date),
      y = `Page Views (Log10)`
  )
```

The prophet package only requires two variables: *ds* and *y*. The *ds* variable is used for the date. The dates should be contiguous and span a minimum of a 2-year time period. The *y* variable represents the outcome you want to forecast which is the log of the *Page Views*.

Note   that you used the log of the page view count instead of the actual page views count. That is a common technique used in situations like you have in this example where you have high data dispersion. Using the log of the page views reduces the dispersion and makes the data easier to visualize. Using the log transformations is also a common technique used in feature engineering when building data science models.

Step 10: Generate the forecast

The three lines of the following code produce the forecast:

```
m <- prophet(dfPageViews)
future <- make_future_dataframe(
            m,
            periods = 365,
            freq = "day",
            include_history = TRUE
        )
forecast <- predict(m, future)
```

The first line produces a forecast model object based on the *dfPageViews* data frame. The second line of code creates a data frame that adds the number of periods specified in the *periods* argument to the end of the dates in the *dfPageViews* data frame. The frequency of the period is defined in the *freq* argument, and if you set the *include_history* argument to *TRUE,* then the historical dates are included in the data frame. The third line generates the forecast via the *predict()* function using the model created in line 1 and the *future* data frame.

Step 11: Generate the plot

You generate four plots in this step. The four plots are

- A plot of the forecast
- A plot that shows the trend in the underlying data

- A plot that shows weekly seasonality

- A plot that shows yearly seasonality

The code used to create the plots is shown here:

```
p1 <- plot(m, forecast) + ggtitle(chartTitle) + theme_few()
p <- prophet_plot_components(m, forecast)
p2 <- p[[1]] + theme_few()
p3 <- p[[2]] + theme_few()
p4 <- p[[3]] + theme_few()
p5 <- grid.arrange(p1, p2, p3, p4, nrow =4)
p5
```

The resulting visual was depicted earlier in Figure 2-15. The visual in Figure 2-15 is valuable because not only do you have a visual of the forecast, but you also have visuals of the components that affect the forecast. You can see how the data is trending with the *trend* plot, how the data typically vary throughout the week with the *weekly* plot, and how the data typically vary throughout the year with the *yearly* plot.

The first thing you need to do is define each plot individually. Here are descriptions of how each chart is individually defined:

- *p1* is used to hold the visual of the actual forecast. It is built using the *plot()* function from the *Prophet* package which, underneath the hood, is built using *ggplot*. Because of that, you are able to apply the *theme_few()* theme to it.

- *p* is used to hold the component charts generated by *Prophet* via the *prophet_plot_components()* function. This function returns a list of *ggplot* charts representing different components of forecast. In this case, it is a trend chart, a weekly seasonality chart, and a yearly seasonality chart.

- *p2*, *p3*, and *p4* hold the three individual component charts produced by *p*. As stated earlier, *p* is a list of *ggplot* charts. You need to subset each chart out of the *p* as a *ggplot* chart. That is accomplished using double brackets, *[[*. When you subset an object from a list in R using single bracket, *[*, the item is returned as a list with one element and not a single object in the data type that the underlying element actually is. So, in this example, *p[1]* will not return a *ggplot* chart, but it will return a list that has one item with that item being a *ggplot* chart. If you want to explicitly return the item in its true data type, then you need to subset with two brackets, *[[,* as you did in the preceding code. Note that because each chart was extracted as a *ggplot* chart, you were able to apply the *theme_few()* theme to each of them.

Now that you have the forecast visual saved as a *ggplot* object and the visual for each forecast components saved as a *ggplot* object, you can combine them into one visual using the *grid.arrange()* function from the *gridExtra* package. This is done is line 6 in the preceding code. You want the charts to be stacked on top of each other. You accomplish that task by setting the *nrow* argument to 4. You control the way the visuals are displayed on the screen using this argument. For instance, if you set that argument to 2, then it will combine the charts in two rows resulting in two charts in each row.

Step 12: Add code to Power BI

The entire script to create the forecast visual is in Listing 2-4. Copy and paste it in *R script editor* in Power BI. The visual will be totally responsive to the Quarterback slicer.

Listing 2-4. The R script that produces the forecast chart

```
library(tidyverse)
library(prophet)
library(ggthemes)
library(gridExtra)
library(lubridate)
library(rlang)

# I needed to load rlang. Look into this package.

currentColumns <- sort(colnames(dataset))
requiredColumns <-
    c("Date", "Page Views", "Page Views (Log10)", "Quarterback")
columnTest <- isTRUE(all.equal(currentColumns, requiredColumns))
```

```
qb <- unique(dataset$Quarterback)

if (length(qb) == 1 & columnTest) {

  chartTitle <- paste0(qb, "'s Wikipedia page views forecast analysis")

  dfPageViews <-
    dataset %>%
    mutate(Date = ymd(Date)) %>%
    rename(ds = Date, y = `Page Views (Log10)`)

  m <- prophet(dfPageViews, yearly.seasonality=TRUE)

  future <- make_future_dataframe(m, periods = 365)

  forecast <- predict(m, future)

  p1 <- plot(m, forecast) + ggtitle(chartTitle) + theme_few()
  p <- prophet_plot_components(m, forecast)

  p2 <- p[[1]] + theme_few()
  p3 <- p[[2]] + theme_few()
  p4 <- p[[3]] + theme_few()
  p5 <- grid.arrange(p1, p2, p3, p4, nrow =4)
  p5

} else {

  plot.new()
  title("The data supplied did not meet the requirements of the chart.")

}
```

Line chart with shade

The line chart is arguably one of the most popular charts used in data visualization. It is perfect for plotting data that spans a period of time. Because it is so widely used, all BI tools natively support them.

In exercise, you will learn how to create a line chart in Power BI using R. This line chart will not be your typical line chart. It will include enhancements that were not available in Power BI as of this writing. The R custom visual that you will create was inspired by a visual created by Hadley Wickham in his book *Ggplot2: Elegant Graphics for Data Analysis* (2nd edition, p. 44). The visual he created in the book used a background shading technique to identify the president that was in office at a given point of time in the chart. You will create an interactive R custom visual in Power BI that will do something similar but with enhancements. The visual that you will create is listed in Figure 2-16.

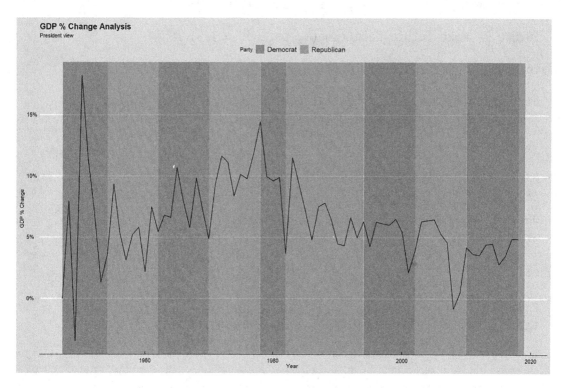

Figure 2-16. *Line chart showing GDP percent change with the background shade based on the political party that reigns during the shaded time*

The R custom visual plots the *% Change of GDP* from calendar year 1947 to calendar year 2018. What's unique about this chart is that the background shade shows the political party of the president at the given time. You leverage the interactivity of Power BI to further enhance the visual. In addition to the president's political party view, you give the end user the ability to change the background to show the party that represented the majority of the United States Senate for a given point of time, the party that represented the majority of the United States House of Representatives for a given point of time, and the party that was the overall majority. In addition to that, you will give the end user the ability to switch between showing the *Actual GDP* and the *% Change in GDP*. The following image in Figure 2-17 shows the overall majority view for Actual GDP.

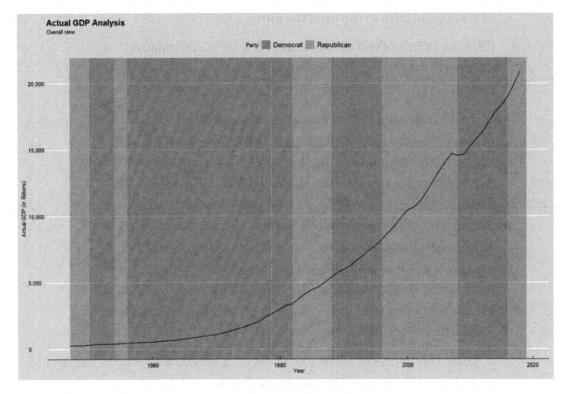

Figure 2-17. *Line chart showing actual GDP with the background shade based on the political party that reigns overall during the shaded time*

Now that you have seen the R custom visual, let's go over the steps you must take to create the visual in Power BI!

Step 1: Acquire the necessary data

1. Go to `https://fred.stlouisfed.org/series/GDP` which will take you to a page on the St. Louis Fed website dedicated to historical GDP information. Click the *EDIT GRAPH* button and change the value in the *Modify frequency* combo box to *Annual* and *Aggregation method* to *End of Period*. Click the *X* in the upper right of the pop-up form to close out the form.

2. Click the *DOWNLOAD* button and choose the *CSV* format. At the time of the writing, that action downloaded a *csv* file to my computer with a file name of *GDP.csv*.

3. Go to the code repo for this example and locate the *PoliticalInfoWithGDP.xlsx* workbook. The *PoliticalInfoWithGDP.xlsx* workbook contains the data sets and the Power Query queries that were used to create the *PoliticalInfoWithGDP* data set used in this example. *PoliticalInfoWithGDP* is the data set that is the source for the visual that will be produced in this exercise. The hard work has been done for you, but you can inspect the workbook to see how the following data sets were combined to produce the *PoliticalInfoWithGDP* data set. Here are descriptions of each query that were developed using Power Query inside the *PoliticalInfoWithGDP.xlsx* workbook:

    a. *GDP*: This query pulls in the GDP data that was downloaded in Step 2 and adds a *% GDP Change* column to the data set.

    b. *Presidents*: This query builds the data set that lists all the US presidents during the time of the analysis along with a column that shows their political affiliation.

    c. *Political Info*: This query gets the information from a Wikipedia article at this URL: `https://en.wikipedia.org/wiki/Party_divisions_of_United_States_Congresses`. The information at this URL is used to get the majority party for the *Senate* and *House of Representative* as well as the president that was in office at the time. *Power Query* is used to reshape the data for the visualization. You can follow the steps in *Power Query* to see how it was done.

Power Query is a data wrangling tool built inside of several Microsoft technologies such as Microsoft Excel, Power BI, Power BI Dataflows, and SQL Server Analysis Services. It has a GUI that greatly reduces that time it takes to perform simple data wrangling. I highly recommend that R and Python developers who are new to the Microsoft ecosystem take a look at this awesome tool!

     d. *PoliticalInfoWithGDP*: Combines the preceding data sets and puts the results in the shape needed for the visualization in this example via *Power Query*.

4. As stated earlier, the *PoliticalInfoWithGDP* data set is the data set that will be used for the visualization and was built using Power Query as previously described. It has been saved as a *csv* file under the name *PoliticalInfoWithGDP.csv*. You can find the file in the code repo for this example.

Step 2: Load the data into Power BI

Load the *PoliticalInfoWithGDP.csv* into Power BI by selecting *Home* ➤ *GetData* ➤ *Text/CSV*, then browse to the location where you saved the file. Click the *Transform Data* button to go to the Power Query editor. Check to make sure that the fields have the following data types:

Column	Data Type
Index	Whole Number
Year	Whole Number
GDP	Decimal Number
% GDP Change	Decimal Number
Senate Majority	Text
House Majority	Text
President's Majority	Text
Overall Majority	Text

Change the columns to the appropriate data type if the columns do not have the preceding data types. The easiest way to change the data type is by right clicking the column in question, then selecting *Change Type*. A list of available data types will appear. Choose the one that applies to the column in question. After you have verified that the data types are valid, click *Close & Apply* to load the data in the Power BI data model.

Go to the table and make sure that the default behavior for *Year* and *Index* is set to *Do Not Summarize*. To do so, select the field in the *Fields* pane, then go to *Modeling* ➤ *Default Summarization* in the *Properties* section, and select *Don't Summarize*.

Step 3: Create the report slicers

The report has two slicers. There is a slicer that is used so that the end user can select the political view (*House, Senate, President,* or *Overall*) and a slicer that is used to select the GDP stat (*Actual GDP* or *GDP % Change*) that the user wants to display. The work in this step has been done for you in the starter *pbix* template which you can find in the code repo for this example.

Step 4: Configure the R visual

Go to the *Visualization pane* and drag an *R visual* to the report canvas. The R visual icon is located in the lower right section of the *Visualization* pane. Resize the visual to the desired size. Add *Year, GetPoliticalLevel, GetPoliticalLevelName, GetGDPStat,* and *GetGDPStatName* to the *Fields* pane which is located beneath the *Visualization* pane. Sometimes Microsoft Power BI defaults to applying an aggregation function to numeric fields. Check to see if it tried to do that with the *Year* field. If there is aggregate function being applied, then remove it by clicking the down arrow of the field in the *Fields* pane and choose *Don't Summarize*.

Step 5: Export data to R Studio for development

Click the 45° arrow that is located on the title bar of the *R script editor*. This action will export the data frame passed to R and the R starter code to *R Studio*. Please refer to Step 4 in the "Callout chart" section if you need a more detailed explanation.

Step 6: Load the required packages to the script

The code needed to load the required packages is listed here:

```
library(tidyverse)
library(ggthemes)
library(scales)
```

The *dplyr* package from *tidyverse* will be used to perform some data wrangling tasks, the *ggplot2* package from *tidyverse* will be used to create the visualization, the economist theme from the *ggthemes* package is used to format the non-data parts of the chart, and a few functions from the *scales* package will be used for number formatting.

Step 7: Create the variables needed for the data validation test

Here is the code for the validation test:

```
currentColumns <- sort(colnames(dataset))
requiredColumns <-
      c("GetGDPStat", "GetGDPStatName", "GetPoliticalLevel",
        "GetPoliticalLevelName", "Year")
columnTest <- isTRUE(all.equal(currentColumns, requiredColumns))

politicalLevelName <- unique(dataset$GetPoliticalLevelName)
gdpStatName <- unique(dataset$GetGDPStatName)
```

The following three tests are performed:

- A test to make sure the data set passed to the visualization contains the necessary fields

- A test to make sure only one political view is in the data set

- A test to make sure only one GDP stat is in the data set

Let's go over the code used in the first test. Here is the code:

```
currentColumns <- sort(colnames(dataset))
requiredColumns <-
    c("GetGDPStat", "GetGDPStatName", "GetPoliticalLevel",
       "GetPoliticalLevelName", "Year")
columnTest <- isTRUE(all.equal(currentColumns, requiredColumns))
```

The first line gets the name of the columns from the data set passed to R and assigned it to a character vector named *currentColumns*. The *sort()* function is used to sort the column names in alphabetical order. The next line of code creates a character vector of the required fields and names it *requiredColumns*. Note that the names are supplied in alphabetical order. That is done to ensure that you can do a direct comparison to the *currentColumns* character vector. The third line of code performs the test by comparing the *currentColumns* character vector to the *requiredColumns* character vector using the *all.equal()* function. When the *currentColumns* is equal to *requiredColumns, TRUE* is returned. Otherwise, information is returned that tries to explain why they did not match. You need a Boolean response so you wrap *all.equal()* with the *isTRUE()* function. Doing so will return *TRUE* when *all.equal()* returns *TRUE*; otherwise, it will return *FALSE*.

Next, two character vectors are created that hold the unique elements in the *GetPoliticalLevelName* field and *GetGDPStatName* field, respectively, using the following code:

```
politicalLevelName <- unique(dataset$GetPoliticalLevelName)
gdpStatName <- unique(dataset$GetGDPStatName)
```

Ideally, both will only have one element, and that will be tested in the next step.

Step 8: Create the data validation test

The template used for the validation is as follows:

```
if(length(politicalLevelName) == 1 &
    length(gdpStatName) == 1 & columnTest) {

    <code to produce visual>

} else {
```

```
plot.new()
title("The data supplied did not meet the requirements of the chart.")

}
```

The preceding *if* statement will execute the code needed to build the visualization if there is only one value in the *GetPoliticalLevelName* field, if there is one value in the *GetGDPStatName* field, and if *columnTest* evaluates to TRUE.

Step 9: Create a new data frame based on the *dataset* data frame

It is best practice not to modify the *dataset* data frame that was passed to R from Power BI. To get around not modifying that data frame, you need to create a new one based on the *dataset* data frame and modify that instead. Here is the code used to create the new data frame:

```
dfPI <- dataset
```

Step 10: Create the variables that will be used for the dynamic portions of the chart

Here's the code needed to define those variables:

```
politicalLevelName <- unique(dfPI$GetPoliticalLevelName)
gdpStatName <- unique(dfPI$GetGDPStatName)

yAxisName <-
  paste(
    gdpStatName,
    ifelse(gdpStatName == "Actual GDP","(in Billions)",""),
    sep = " "
  )
chartTitle <- paste(gdpStatName, "Analysis", sep = " ")
chartSubtitle <- paste(politicalLevelName, "view", sep = " ")
```

The preceding code dynamically produces the label for the *y* axis, the chart title, and the chart subtitle. The first two lines get the *political level* and the *GDP stat* and store the information for later use. The next line of code, actually six lines in the preceding code snippet because of formatting reasons, is used to create the label for the *y* axis. It uses the *paste()* function to combine the value of the *gdpStatName* variable with the string *(in billions)* if the GDP stat that is being used is *Actual GDP*. The result is assigned to the *yAxisName* variable. The fourth line of code dynamically builds the main chart title by concatenating the value of the *gdpStatName* variable and the word "Analysis" using a space delimiter. The last line of code dynamically creates the chart subtitle by concatenating the value of the *politicalLevelName* variable and the word *view* using a space delimiter.

Step 11: Create the data sets needed for background shade

In this step, you will use the data in the *dfPI* data frame and transform it into the data set you need for the background shade portion of the visualization. Let's start by taking a peek at the data frame in Figure 2-18.

Year	GetPoliticalLevel	GetPoliticalLevelName	GetGDPStat	GetGDPStatName
1947	Republican	Overall	259.745	Actual GDP
1948	Republican	Overall	280.366	Actual GDP
1949	Republican	Overall	270.627	Actual GDP
1950	Democrat	Overall	319.945	Actual GDP
1951	Democrat	Overall	356.178	Actual GDP
1952	Democrat	Overall	380.812	Actual GDP
1953	Democrat	Overall	385.970	Actual GDP
1954	Republican	Overall	399.734	Actual GDP
1955	Republican	Overall	437.092	Actual GDP

Figure 2-18. *A peek of the dfPI data frame*

You need to use the preceding data to create a data set that has the start time and end time of each contiguous party reign period. You need to take the preceding data frame and write code that will transform it to the data frame in Figure 2-19.

group_id	Party	start	end
1	Republican	1947	1949.99
2	Democrat	1950	1953.99
3	Republican	1954	1955.99
4	Democrat	1956	1981.99
5	Republican	1982	1987.99
6	Democrat	1988	1995.99
7	Republican	1996	2007.99
8	Democrat	2008	2015.99
9	Republican	2016	2018.99

Figure 2-19. *Required data frame format*

In order to accomplish this transformation, you need to be able to uniquely identify each contiguous political party reign. Currently, there is no unique identifier for each political party reign so you need to add one. Once you do so, you will be able to use traditional aggregation methods to get the start and end year of each reign. Let's first go over the steps needed to add the unique identifier which you will call *group_id*. The code that produces the information for the *group_id* is listed here:

```
dfPI$GetPoliticalLevel <- as.character(dfPI$GetPoliticalLevel)
runs <- rle(dfPI$GetPoliticalLevel)
group_id <- rep(seq_along(runs$lengths), runs$lengths)
```

The first thing that is done in line 1 is a data type conversion. The default data type for columns in a data frame that contains strings in R is a data type known as *factor* for R versions prior to R 4.0.

As of the writing of this book, the Power BI service has not been updated to R version 4.0. If that is the case at the point of time you are working through this exercise, then you will also need to do the conversion. If the Power BI service has been updated to R version 4.0 or later, then you will not have to do the conversion step.

Since the data in the *GetPoliticalLevel* column are all strings, R has converted it to the factor data type. In order to perform the data processing needed in this step, you need to convert the *GetPoliticalLevel* to a character data type. That data type is comparable to the *Text* data type in DAX. That is accomplished in the first line of code. The next thing you do is use the *rle()* function from base R to define the groups in the *GetPoliticalLevel* column. The "rle" in the *rle()* function stands for *run length encoding*. When you pass the *GetPoliticalLevel* column to the *rle()* function, you get the following output illustrated in Figure 2-20.

```
Run Length Encoding
  lengths: int [1:9] 3 4 2 26 6 8 12 8 3
  values : chr [1:9] "Republican" "Democrat" "Republican" "Democrat" "Republican" "Democrat" "Republican"
 ..:.
```

Figure 2-20. *Output from the rle() function*

The *rle()* function returned a list that contains two vectors. The first vector is a vector named *lengths* and, in this example, it contains the length of each political reign. The second vector is named *values* and, in this example, it contains the party who was in power in each political reign. The information returned from the *rle()* function is saved in the *runs* variable.

The information contained in the *runs* variable is used to create the *group_id* vector using the following code:

```
group_id <- rep(seq_along(runs$lengths), runs$lengths)
```

The first thing that is done is the *seq_along()* function is used to generate a sequence of integers that will represent the *group_id*. Since *runs$length* has a length of 9, an integer vector will be returned that contains numbers 1–9. That expression is put into the *rep()* function as the first argument, and the vector *runs$length* is put into the second argument. Since both arguments contain vectors with the same length, the rep function will take the first element produced by the *seq_along()* function, which in this case is the number 1, and repeat it the number of times represented in the second integer vector, which in this case is 3. So, the number 1 will be repeated three times. Next, the number 2 will be repeated four times because the value of the second element in *runs$lengths* is 4. That logic is repeated for each element in *runs$lengths,* and the resulting output is listed in Figure 2-21.

```
[1] 1 1 1 2 2 2 2 3 3 4 4 4 4 4 4 4 4 4 4 4 4 4 4 4 4 4 4 4 4 4 4 5 5 5 5 5 6 6 6 6 6 6 6 7
[51] 7 7 7 7 7 7 7 7 7 7 7 8 8 8 8 8 8 8 8 9 9 9
```

Figure 2-21. *The output of the group_id variable*

Next, you use the following code to create the data set for the visualization:

```
dfShadeInfo <-
  cbind(dfPI, group_id) %>%
  transmute(group_id, Year, Party = GetPoliticalLevel) %>%
  group_by(group_id, Party) %>%
  summarize(start = min(Year), end = max(Year)+0.99) %>%
  ungroup()
```

The preceding code starts off by appending the *group_id* vector as a new column to the *dfPI* data frame using the *cbind()* function where the *c* in *cbind()* stands for column. Next, the pipe, *%>%*, operator is used to pass the resulting data frame as the first argument to the *transmute()* function in the next line.

The *transmute()* verb is like the *mutate()* verb from *dplyr* in that it is used to project new columns, but the difference with the *transmute()* function is that the data frame that it creates is based only on the columns defined in the function.

The preceding *transmute()* verb will produce a data frame with three columns named *group_id, Year,* and *Party*. Figure 2-22 is a peek of the first nine rows to show what the result of using the *transmute()* verb in the preceding code looks like.

group_id	Year	Party
1	1947	Republican
1	1948	Republican
1	1949	Republican
2	1950	Democrat
2	1951	Democrat
2	1952	Democrat
2	1953	Democrat
3	1954	Republican
3	1955	Republican

Figure 2-22. *The result of using the transmute() verb*

Note that now you have a unique identifier for each political reign! Now you can use traditional aggregation functions to get the start and end time for each political reign. Again, the pipe operator is used to pass the data frame created by the *transmute()* verb as the first argument of the *group_by()* verb in the next line. The *group_by()* verb will group the data frame passed to it by the *group_id* and *Party* columns. The result of the *group_by()* will be passed as the first argument of the *summarize()* verb in the next line. The *summarize()* verb enables you to apply aggregation functions to your data set based on your groupings. So, you are able to create a column named *start* that represents the start of the political party reign based on the minimum year in the grouping, and you are able to create a column named *end* that represents the end of the political party reign based on the maximum year in the grouping. The value 0.99 is added to *end* column to minimize the gap between the start of the old reign and beginning of the new reign.

Step 12: Create the data sets needed for line chart

The code used to create the data set for the line chart is as follows:

```
dfLineChartInfo <-
  dfPI %>%
  transmute(Year, GetGDPStat)
```

The goal is to develop a line chart that displays the chosen GDP statistic by year that spans the time period of the data set. In order to do that, we need to supply the data set with two columns. Those columns are *Year* and *GetGDPStat*. You create the needed data set by extracting those columns from the *dfPI* data frame using the *transmute()* function.

Step 13: Create a named character vector that will be used to color the shades

```
partyColors = c("Republican"="red", "Democrat"="blue",
                "Tie" = "white")
```

The color for Republican party is red, and the color for the Democratic party is blue. You can force the visualization to use those colors for the parties via the named character vector *partyColors*. The *partyColors* named character vector will be used later in the script to force the visual to shade the background *red* for the time periods when the republicans were in control and will force the visual to shade the background *blue* for the time periods when the democrats were in control.

Step 14: Start the chart by defining the ggplot function

The previous steps focused on data prep. The data prep steps illustrated one of the nice features about R visuals in Power BI. If the data set that is passed to your visual is not in the proper shape, you can use R to reshape your data set to get it where it needs to be.

In this step, you initiate the creation of the chart by setting the *ggplot()* function using the following code:

```
p <- ggplot()
```

You might have noticed that you are initiating your chart with a *ggplot()* function with any arguments. You did so because your chart will contain two layers, each with independent data sources and aesthetics so they will be set at the layer level.

Step 15: Add a layer to create the background shade

This is accomplished using the following code:

```
geom_rect(
    data = dfShadeInfo,
    aes(xmin = start, xmax = end, fill = Party),
        ymin = -Inf, ymax = Inf, alpha = 0.4, color = NA
    )
```

The source data for this layer is the *dfShadeInfo* data frame. Figure 2-23 shows how it looks if the *Overall* political view is chosen.

group_id	Party	start	end
1	Republican	1947	1949.99
2	Democrat	1950	1953.99
3	Republican	1954	1955.99
4	Democrat	1956	1981.99
5	Republican	1982	1987.99
6	Democrat	1988	1995.99
7	Republican	1996	2007.99
8	Democrat	2008	2015.99
9	Republican	2016	2018.99

Figure 2-23. *The dfShadeInfo data frame*

The preceding image shows nine political reigns, so the *geom_rect()* function will use the preceding data frame to shade the background using a color based on the political party that was in power during each of the political reign. The start and end point of each political reign is defined by setting the *start* column to the *xmin* argument in the *aes()* function and setting the *end* column to the *xmax* argument in the *aes()* function. The upper boundary is defined by setting the *ymax* argument to *Inf,* and the lower boundary is defined by setting *ymin* to *-Inf*. Using infinity for the upward and lower boundaries ensures that the shade will be vertically filled. The shade is painted for each reign, and the result is the shaded background listed in Figure 2-24.

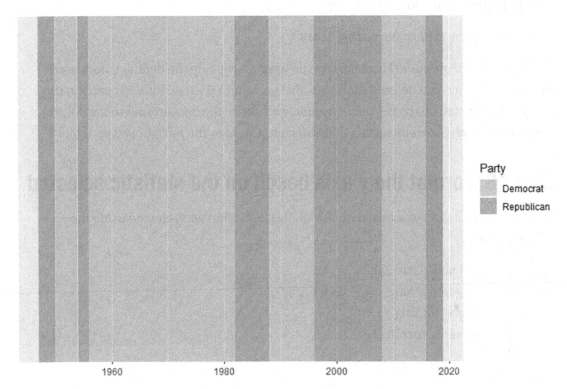

Figure 2-24. *Background shade produced by the geom_rect() geom*

Note the shading corresponds to the time period of each reign. The colors used for each political party will be changed later in the script.

Step 16: Add a line chart based on the statistic selected

The following code is used to create the line chart layer of the visualization:

```
geom_line(data = dfLineChartInfo, aes(x = Year, y = GetGDPStat))
```

The *geom_line()* geom is used to create a line chart. The data frame that it is based on is the *dfLineChartInfo* data frame, and the column in that data frame that is used for the x axis is the *Year* column and the column that is used for the y axis is the *GetGDPStat* column.

Step 17: Reshade the background using pre-determined colors based on the political party

That is done using the following code:

```
scale_fill_manual(values=partyColors)
```

The *scale_fill_manual()* function can be used to override the default colors used by *ggplot2*. Earlier, you defined the colors that you wanted to use for each party in the *partyColors* named character vector. *ggplot2* will shade the background to the colors specified in *partyColors* by setting the *values* argument to the *partyColors* variable.

Step 18: Format the y axis based on the statistic selected

Next, you format the *y* axis based on the GDP statistic that was selected using the following code:

```
scale_y_continuous(labels=
    ifelse(gdpStatName == "ActualGDP",
        comma_format(),
        percent_format()
    )
) +
```

If the selection was made to return the actual GDP, then the *y* axis will be formatted in a *comma_format()* because the *ActualGDP* needs to be formatted as a number. Otherwise, the *y* axis will be formatted as a percent using *percent_format()* because the chart will be displaying percent change of GDP. Both the *comma_format()* and the *percent_format()* function come from the *scales* package.

Step 19: Add labels to the x and y axis

The following code adds a dynamically created label to the *y* axis and a fixed label to the *x* axis:

```
    ylab(yAxisName) +
    xlab("Year")
```

Step 20: Add the dynamic titles and caption to the custom R visuals

The *ggtitle()* function is used to dynamically create the chart title and the chart subtitle using the following code:

```
ggtitle(chartTitle, subtitle = chartSubtitle)
```

Step 21: Apply a theme based on *The Economists* publication

Since you are visualizing economic data, you will use the *theme_economist()* theme from the *ggthemes* package to format your visual. The *theme_economist()* theme uses formatting rules that are used by *The Economists* publication. You apply the *theme_economist()* theme to your visual by adding the line of this code to the script:

```
theme_economist()
```

Step 22: Add code to Power BI

The entire script that produces the visualization in this exercise is listed in Listing 2-5. The script is functional so you can copy and paste it into Power BI. Make sure to paste it after the commented section in the Power BI *R script editor*.

Listing 2-5. The R script that produces the line chart with shade chart

```
library(tidyverse)
library(ggthemes)
library(scales)

currentColumns <- sort(colnames(dataset))
requiredColumns <- c("GetGDPStat", "GetGDPStatName", "GetPoliticalLevel",
"GetPoliticalLevelName", "Year")
columnTest <- isTRUE(all.equal(currentColumns, requiredColumns))

politicalLevelName <- unique(dataset$GetPoliticalLevelName)
gdpStatName <- unique(dataset$GetGDPStatName)
```

```
if(length(politicalLevelName) == 1 & length(gdpStatName) == 1 & columnTest)
{
  dfPI <- dataset

  #Variables for dynamic portions of the chart
  politicalLevelName <- unique(dfPI$GetPoliticalLevelName)
  gdpStatName <- unique(dfPI$GetGDPStatName)
  yAxisName <- paste(gdpStatName, ifelse(gdpStatName == "Actual GDP",
  "(in Billions)",""), sep = " ")
  chartTitle <- paste(gdpStatName, "Analysis", sep = " ")
  chartSubtitle <- paste(politicalLevelName, "view", sep = " ")

  #Get the number of
  dfPI$GetPoliticalLevel <- as.character(dfPI$GetPoliticalLevel)
  runs <- rle(dfPI$GetPoliticalLevel)
  group_id <- rep(seq_along(runs$lengths), runs$lengths)

  #Create the data set for shade layer
  dfShadeInfo <-
    cbind(dfPI, group_id) %>%
    transmute(group_id, Year, Party = GetPoliticalLevel) %>%
    group_by(group_id, Party) %>%
    summarize(start = min(Year), end = max(Year)+0.99) %>%
    ungroup()

  #Create the data set for line chart layer
  dfLineChartInfo <-
    dfPI %>%
    transmute(Year, GetGDPStat)

  #Define the shade colors
  partyColors = c("Republican"="red", "Democrat"="blue", "Tie" = "white")

  #Create the visualization
  p <- ggplot() +
      geom_rect(
          data = dfShadeInfo,
          aes(xmin = start, xmax = end, fill = Party),
              ymin = -Inf, ymax = Inf, alpha = 0.4, color = NA
          ) +
```

```
    geom_line(data = dfLineChartInfo, aes(x = Year, y = GetGDPStat)) +
    scale_fill_manual(values=partyColors) +
    scale_y_continuous(
      labels=ifelse(gdpStatName == "Actual GDP", comma_format(),percent_
      format())
    ) +
    ylab(yAxisName) +
    xlab("Year") +
    ggtitle(chartTitle, subtitle = chartSubtitle) +
    theme_economist()

  p
} else {

  plot.new()
  title("The data supplied did not meet the requirements of the chart.")

}
```

Map

Spatial analytics has become more important in recent years. One form of spatial analytics that is popular in business is *geographic-based spatial analytics*. Power BI offers built-in map capabilities that work great for most situations. But in this example, I will show how you can use R to build custom maps. The visual that you will draw is a *state heat map* that shades the counties in the state based on the population quintile the county falls in. The darker the shade, the more populated the county is. Figure 2-25 is an example of the visual you will create when the state of Indiana is selected.

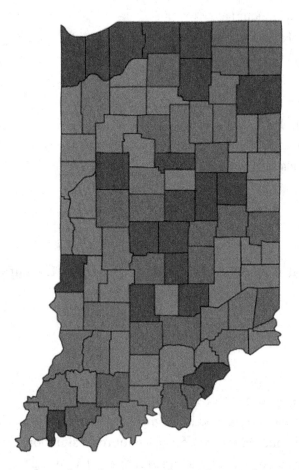

Indiana's County Population Analysis
(the darker shades the higher the population)

Figure 2-25. *Map visual produced using ggplot2 when Indiana is selected*

The visual was created using a data set that contains the coordinates that define the borders of the counties in the continental United States. When a state is selected, *ggplot2* uses the geographic coordinates to draw the map. The counties are filled with a shade of blue that is darker in the more populated counties and a shade of blue that is lighter in the lower populated counties. You can easily switch between states by leveraging the interactive capabilities of Power BI. If you change the slicer from Indiana to Kentucky, then the following visual in Figure 2-26 will appear.

Kentucky's County Population Analysis
(the darker shades the higher the population)

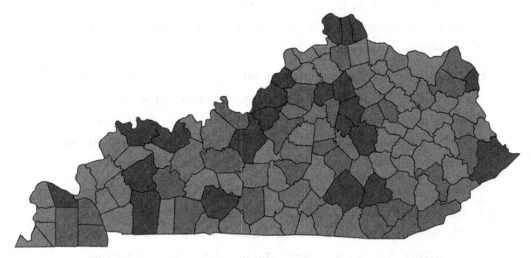

Figure 2-26. *Map visual produced using ggplot2 when Kentucky is selected*

One of the features that makes this visual great is that it limits the geography to only the selected geography. Thus, it does not waste valuable chart real estate on other geographies. Many map visuals often includes other geographical areas in addition to the one that is the point of interest. Also, because it is created using *ggplot2*, you can add other dynamic features such as a dynamic *chart title*, a dynamic *chart subtitle*, and a dynamic *chart caption*. Let's go over the steps needed to create this amazing visual!

Step 1: Acquire the necessary data

1. Obtain a data set that lists the states with both their full name and abbreviated name. The data set used came from https://en.wikipedia.org/wiki/List_of_U.S._state_abbreviations and can be found in the *Abbreviations* tab in the *DataWrangling.xlsx* workbook.

2. Build out a data set with the county information needed for the visual. Information used to create the data set was obtained from www.nrcs.usda.gov/wps/portal/nrcs/detail/national/home/?cid=nrcs143_013697 and can be found in the *County FIPS Codes* tab of the *DataWrangling.xlsx* workbook.

3. Obtain a data set from the Census that lists population by county.
   The data set can be found in the *PopulationInfo* tab in the
   DataWrangling.xlsx workbook along with the Python script used
   to obtain the data set.

4. Create a data set that contains the missing *FIPS code*. Some of the
   missing *FIPS codes* were due to mapping issues, and some were
   because they were not in the *County FIPS Codes* data set. This
   data set will be used to correct that problem. It can be found in the
   TheMissingInfo worksheet in the *DataWrangling.xlsx* workbook.

5. Obtain a data set from the *ggplot2* package that contains border
   information for the counties in the continental United States.
   The information is warehoused in the *Map Info* tab of the
   DataWrangling.xlsx workbook, and here is the code used to get
   the data set:

```
library(tidyverse)
us_counties <- map_data("county", ".")
path <- "<location where you want to save the data set>"
write_csv(us_counties, path)
```

6. Use *Power Query* to combine the information from the previous
   data sets into a shape needed by the visualization. The data set
   can be found in the *ChartData* tab in the *DataWrangling.xlsx*
   workbook. The query used to create the data set is also named
   ChartData, and you can look at the query to see the steps used to
   create the data set. This is the data set that will be used by Power
   BI to create the R visual.

7. Save the data in the *ChartData* tab of the workbook. The data was
   also saved to a *csv* file named *chartdata.csv*.

As with most visualization projects and data science projects in general, the longest
part of the project is normally acquiring, cleaning, and shaping the data. In this exercise,
I did the hard work for you so that you can focus on the data visualization aspect. You
can review the material in the code repo for this exercise to see the steps I took to
wrangle the data. For those of you coming from an R and/or Python background, doing
so will be a good way to familiarize yourself with *Power Query*.

Step 2: Load the data into Power BI

You need to bring the *chartdata.csv* file into the Power BI, then make some minor adjustments to the data set before you load it into the data model. You can refer to the *pbix* file for this example to see the transformations.

Step 3: Create a slicer based on state in the Filter pane

Create a slicer based on the *State* column by first dragging the *State* column to the report canvas. After you do that, you need to change the visual to a slicer by changing the visualization type to a *slicer* visual. The *slicer* is the visual with the icon that has an image of a funnel in it.

Step 4: Configure the R visual

Go to the visualization pane and drag an R visual to the report canvas. Resize the visual to the desired size. Drop the *index, lat, long, County, Total Population,* and *State* fields from the *chartdata* table into the *Fields* pane. If the Fields pane is not showing, select the R visual and it should appear.

Step 5: Export data to R Studio for development

Click the 45° arrow that is located on the title bar of the *R script editor*. This action will export the data frame passed to R and the R starter code to *R Studio*. Please refer to Step 4 in the "Callout chart" section if you need a more detailed explanation.

Step 6: Load the required packages

The packages that will be used for this script are listed here:

```
library(tidyverse)
library(ggthemes)
```

The *dplyr* package from *tidyverse* will be used for some data wrangling, the *stringr* package from *tidyverse* is used to reformat some titles, and the *ggplot2* package from *tidyverse* will be used to create the map visual. The *theme_map()* theme from the *ggthemes* package will be used to format the non-data elements of the map visual.

Step 7: Create the variables needed for the data validation test

Here is the code that creates the variables for the validation test:

```
currentColumns <- sort(colnames(dataset))
requiredColumns <- c("County", "Index", "lat", "long", "State", "Total
Population")
columnTest <- isTRUE(all.equal(currentColumns, requiredColumns))
state <- str_to_title(unique(dataset$State))
```

The first line gets the column names from the data set passed to R from Power BI and sorts them in alphabetical order, then put the results in the *currentColumns* character vector variable. The second line creates a character vector that contains the column names that are expected by the visualization in alphabetical order and assigns it to a character vector named *requiredColumns*. The third line performs a test comparing the column names in the *currentColumns* character vector to the column names in the *requiredColumns* character vector. If they match, then *TRUE* is returned; otherwise, *FALSE* is returned. The fourth line gets all of the unique elements in the *State* column of the *dataset* data frame and uses the *str_to_title()* function from the *stringr* packages to put the state names in a proper name format.

Step 8: Create the data validation test

The shell of the data validation is as follows:

```
if (length(state) == 1 & columnTest) {

   <execute code to build visual>

} else {
  plot.new()
  title("The data supplied did not meet the requirements of the
        chart.")
}
```

The source data frame of the visualization needs to only contain one state, and it needs to contain the required columns. The preceding *if* statement tests to see if there is only one state in the data frame and if the data frame passed the column test. If both conditions are met, then it will proceed to execute the code to create the map visual; otherwise, it will produce a blank visual with a message informing you that R did not have adequate information and could not create the visual.

Step 9: Create the variables for the chart titles

Here is the code needed to create the chart titles:

```
chartTitle <- paste0(state, "'s County Population Analysis")
subTitle <- "(the darker shades the higher the population)"
```

The main chart title is made by concatenating the selected state with the string "'s County Population Analysis" and is assigned to the *chartTitle* variable. The string "(the darker shades the higher the population)" is used for the subtitle, and it is assigned to the *subTitle* variable.

Step 10: Add the quintile column to the data set

If you recall, the goal of this visualization is to group the counties in each state by quintiles based on population. You use the *ntile()* function from *dplyr* to group the counties in the state in quintiles as illustrated in the following code:

```
chartdata <-
    dataset %>%
    mutate(quintile = factor(ntile(`Total Population`, 5)))
```

The code creates a new data frame called *chartdata* by starting with the *dataset* data frame, then the pipe operator from *dplyr* is used to pass it as the first argument to the *mutate()* verb in the next line. The *mutate()* verb is used to project a new column named *quintile* that will be created using the *ntile()* function. The *ntile()* function is wrapped with the factor function to convert the quintiles to a factor data type. This action has a similar effect as *Don't Summarize* does in Power BI. Converting the quintiles to factor causes them to be treated as categorical data and not as numeric data.

Step 11: Define the colors that will be used to shade the map

You are going to use different shades of blue to shade the counties in the selected state. You are going to use five different shades of blue using the name character vector listed here:

```
quintileColors <-
  c(
    "1" = "dodgerblue",
    "2" = "dodgerblue1",
    "3" = "dodgerblue2",
    "4" = "dodgerblue3",
    "5" = "dodgerblue4"
  )
```

The *dodgerblue* color listed for *quintile 1* is the lightest shade of blue in the list, and *doderblue4* for *quintile 5* is the darkest shade of blue in the list. Here, you are using the color names to define them instead of the hexadecimal value. Go to the following website to learn more about the colors available to you in *ggplot2*: www.stat.columbia. edu/~tzheng/files/Rcolor.pdf. You will see available names you can use to refer to specific colors.

Step 12: Define the ggplot() function

You start by defining the *ggplot()* function using the following code:

```
ggplot(chartdata, aes(long, lat, group = County,
      fill = quintile))
```

You use the *chartdata* data frame as the chart's data source. Next, you configure the *aes()* function by setting the *long* column to the *x* coordinate, the *lat* column to the *y* coordinate, the *County* to the *group* aesthetic, and the *quintile* to the *fill* aesthetic. This is our first time using the *group* aesthetic because prior to this example, *we* have been using *individual geoms,* but the *geom_polygon()* geom is what is known as a *collective geom.*

Individual geoms only depend on one row, but *collective* geoms depend on multiple rows of data. This example will use the *geom_polygon()* geom which is a *collective* geom, and it will use data from multiple rows to define the border of the polygon visual that will be created.

The *geom_polygon()* in this example draws polygons using the latitude and longitude points in the data set for each *group* with the group in this example being the *County* in the data set. The *fill* property is used to tell *ggplot2* how you want to color the inside of each county. Here, you will color the inside of each County based on the quintile they are in.

You were able to use a *Cartesian coordinate system* (in this example, a coordinate system that is based on an x and y axis) to represent your spatial data because you are looking at a relatively small area of the globe. Even though the earth is the shape of a sphere, when you are looking at a relatively small geographical area you can analyze it using the same techniques that are used to analyze a plane and get relatively good results.

Step 13: Add the map layer

Here is the code that adds the map layer:

```
geom_polygon(show.legend = FALSE, color = "black")
```

You already have the aesthetics that you need for the map defined in the *ggplot()* function. They will carry over to the *geom_polygon()* geom because anything that is defined in the *ggplot()* function is available to all subsequent layers. You don't need a legend in this visual, so you set the *show.legend* argument to *FALSE*. You want the color of the borders to be black. You make them black by setting the color argument to black.

Note   that when you want to set a chart attribute to a specific color, you need to do so without using the *aes()* function. The reason was explained in detail in Chapter 1.

117

Step 14: Format the x and y axis

The following code is used to reformat the *x* and *y* axis:

```
scale_x_continuous(name = NULL, labels = NULL, breaks = NULL) +
scale_y_continuous(name = NULL, labels = NULL, breaks = NULL)
```

You removed the name, labels, and breaks by setting them equal to NULL. You did so because you are visualizing a map and that information is not needed.

Step 15: Color the counties based on their quintile

The following code will be used to color the counties:

```
scale_fill_manual(values = quintileColors)
```

The *ggplot2* package has good defaults which is one of the reasons that makes the package so nice. What makes the package even nicer is that you can easily override the defaults. The map will be colored as depicted in Figure 2-27 if the defaults from *ggplot2* are used.

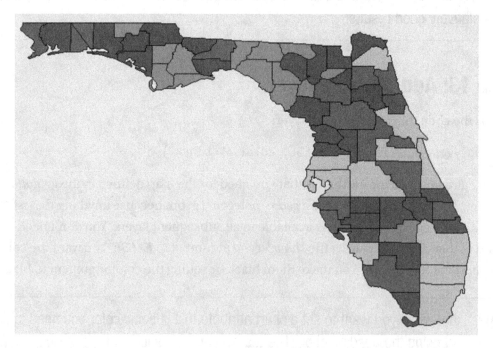

Figure 2-27. *The map visual with the default colors that is produced when Florida is selected*

Figure 2-28 is what the map looks like after you use custom colors via the *scale_fill_manual()* function.

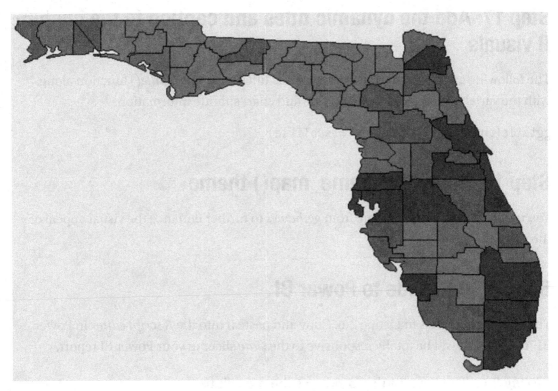

Figure 2-28. *The map visual with custom colors that is produced when Florida is selected*

The second version makes it easy for the report user to tell which counties have the lower populations and which counties have the higher populations based on the darkness of the shades.

Step 16: Improve the approximation of the selected state

The following code is used to do a quick aspect ratio approximation based on the *mercator map* projection:

```
coord_quickmap()
```

The *geom_polygon()* does a good job creating the map, but using the preceding code quickly makes the map display a little more accurately.

Step 17: Add the dynamic titles and caption to the custom R visuals

The following code is used to add the dynamic titles using the *ggtitle()* function along with the variables that hold the chart title and chart subtitle information:

```
ggtitle(chartTitle, subtitle = subTitle)
```

Step 18: Apply the theme_map() theme

You use the *theme_map()* theme from *ggthemes* to further enhance the visual appeal of the map.

Step 19: Add code to Power BI

The complete script is in Listing 2-6. Copy and paste it into the *R script editor* in Power BI. The R visual will be totally responsive to the *state* slicer in your Power BI report.

Listing 2-6. The R script that produces the map chart

```
library(tidyverse)
library(ggthemes)

currentColumns <- sort(colnames(dataset))

requiredColumns <-
   c("County", "Index", "lat", "long", "State",
     "Total Population")

columnTest <- isTRUE(all.equal(currentColumns, requiredColumns))
state <- str_to_title(unique(dataset$State))

if (length(state) == 1 & columnTest) {

  chartTitle <- paste0(state, "'s County Population Analysis")
  subTitle <- "(the darker shades the higher the population)"
```

```
chartdata <-
  dataset %>%
  mutate(quintile = factor(ntile(`Total Population`, 5)))

quintileColors <-
  c(
    "1" = "dodgerblue",
    "2" = "dodgerblue1",
    "3" = "dodgerblue2",
    "4" = "dodgerblue3",
    "5" = "dodgerblue4"
  )

ggplot(chartdata, aes(long, lat, group = County,
        fill = quintile)) +
  geom_polygon(show.legend = FALSE, color = "black") +
  scale_x_continuous(name=NULL, labels=NULL, breaks=NULL) +
  scale_y_continuous(name=NULL, labels=NULL, breaks=NULL) +
  scale_fill_manual(values = quintileColors) +
  coord_quickmap() +
  ggtitle(chartTitle, subtitle = subTitle) +
  theme_map()
} else {
  plot.new()
  title("The data supplied did not meet the requirements of the
        chart.")

}
```

Quad chart

As a Microsoft data professional, you are well aware of the visual in Figure 2-29 produced by Gartner.

Figure 2-29. *The 2020 Gartner Magic Quadrant Chart*

Microsoft has done well compared to its competitors in the *Business Intelligence* space, and Gartner acknowledges it. Microsoft has used industry comparisons like the *Gartner Magic Quadrant Chart* to tout their position in the industry.

The visual Gartner used, the *quad chart*, is very effective when comparing different entities based on two metrics. In this case, Gartner is comparing different BI vendors based on their *Ability to Execute* and their *Completeness of Vision*. Gartner made the chart visibly appealing by adding minor details. Some of those details include

- Labeling the quadrants

- Adding background shade to quadrants 2 and 3

- Using non-traditional labels for the x and y axis

- Adding a chart caption

These subtleties are not easy to create in Power BI but are straightforward in R using *ggplot2*.

In this section, I will illustrate my point using a data set based on "play-by-play" data for the 2008–2009 LA Lakers season. That info will be used to create the quad chart in Figure 2-30 that compares the players against each other based on *total points scored* and *total rebounds made*.

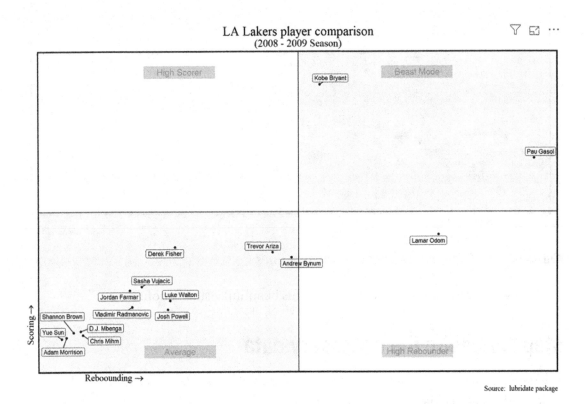

Source: lubridate package

Figure 2-30. *Quad chart comparing LA Lakers players based on rebounds and points scored*

Like in the previous visualizations, you will take advantage of Power BI's interactive capabilities and add slicers to the visual so that you can see how the situation plays out in different scenarios. There will be a filter for the type of game (home or away) as well as a filter for the period of the game. The visual in Figure 2-31 shows the comparison when only away games are considered.

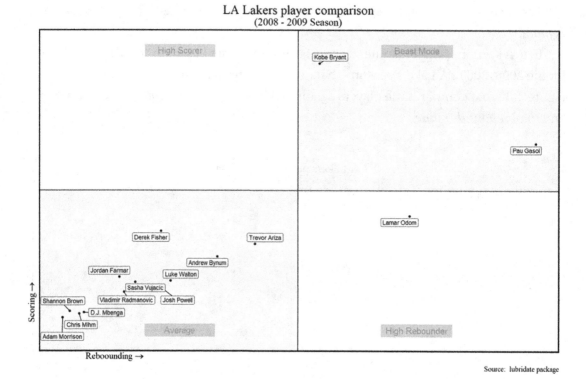

Figure 2-31. *Quad chart based on away games*

Let's go over the steps needed to create this beautiful visualization!

Step 1: Acquire the necessary data

The data set used in this example was obtained from the *lubridate* package in R. The code used to get that data set is as follows:

```
library(tidyverse)
library(lubridate)

chartData <-
  lakers %>%
  filter(
    (player != "" & team == "LAL" &
     result == "made" &
    etype %in% c("shot","free throw")
    ) |
```

```
    (player != "" &
     team == "LAL" &
     etype == "rebound")
    ) %>%
  mutate(
    rebound = ifelse(etype == "rebound",1,0)
  ) %>%
  group_by(game_type, period, player) %>%
  summarize(
    `Total Points` = sum(points),
    Rebounds = sum(rebound)
  ) %>%
  rename(
    `Game Type` = game_type,
    Period = period,
    Player = player
  )

path <- "<path to location to save file>"
write_csv(chartdata, path)
```

The output was saved to a *csv* file because that format is an easy format for Power BI to consume. Don't worry if you don't understand the preceding code because you will learn the techniques used later in the book. For now, know that if you need to recreate the data set used in this visualization, you can do so by running the preceding R script.

Step 2: Load the data into Power BI

Load the data in Power BI from the *csv* file created in Step 1 using *GetData*. Once loaded, the data type of each column should have the data types listed below:

Column Name	Data Type
Game Type	Text
Period	Whole Number
Player	Text
Total Points	Whole Number
Rebounds	Whole Number

The file is already loaded in the *pbix* file located in the folder for this example in the code repo. You can review the Power Query steps to see what transformations were applied.

Step 3: Create a slicer for game type and period

Create a report slicer based on the *Game Type* field by first dragging the *Game Type* field to the report canvas, then selecting the *Slicer* visual in the *Visualization* pane. Make sure the field you added to the canvas is highlighted before you select the *Slicer* visual to convert it into a slicer. The slicer visual is the one that includes a picture of a filter in the image. Perform the same action with the *Period* field.

Step 4: Configure an R visual on the report canvas

Go to the visualization pane and drag an R visual to the report canvas. Resize the visual to the desired size. Drop the *Player*, *Rebounds*, and *Total Points* fields from the *chartData* table into the *Fields* pane. If the Fields pane is not showing, select the R visual and it should appear.

Step 5: Export data to R Studio for development

Click the 45° arrow that is located on the title bar of the *R script editor*. This action will export the data frame passed to R and the R starter code to *R Studio*. Please refer to Step 4 in the *"Callout chart"* section if you need a more detailed explanation.

Step 6: Load the required packages

The required packages for this script are listed here:

```
library(tidyverse)
library(ggrepel)
library(ggthemes)
library(scales)
```

The *dplyr* package from *tidyverse* is used to perform some data wrangling tasks, *ggplot2* from *tidyverse* is used to create the visualization, the *ggrepel* package is used to add the labels to the data points in the visualization, the *theme_tufte()* theme from the *ggthemes* package is used to format the non-data elements of the visual, and the *rescale()* function from the *scales* package is used to rescale some of the data.

Step 7: Create the variables needed for the data validation test

The variables needed for the validation test are defined in the following code:

```
currentColumns <- sort(colnames(dataset))
requiredColumns <- c("Player", "Rebounds", "Total Points")
columnTest <- isTRUE(all.equal(currentColumns, requiredColumns))
noPlayerDups <-
    length(unique(dataset$Player)) == length(dataset$Player)
```

Here is a description of what each line in the preceding code snippet is doing:

- The first line of code creates a character vector variable named *currentColumns* based on the column names from the *dataset* data frame sorted in alphabetical order.

- The second line of code creates a character vector variable named *requiredColumns* that contains the required columns needed for the Quad Chart in alphabetical order.

- The third line of code uses the *all.equal()* function from base R to test the *currentColumns* variable and the *requiredColumns* variable for equivalency. If they are equal, then *all.equal()* returns *TRUE;* otherwise, it returns information about why they are not equal. We wrap *all.equal()* with *isTRUE()* to get a Boolean response.

- The fourth line of code (actually two due to formatting) uses the *noPlayerDups* variable to hold the result of the test that checks if there are any duplicate players in the data set. It does so by comparing the number of unique elements in the *Players* column to the length of the *Players* column. If they are equal, then none of the players are duplicated.

Step 8: Create the data validation test

The template for the validation test is listed here:

```
if (noPlayerDups & columnTest) {

    <code to produce visual>

} else {

  plot.new()
  title("The data supplied did not meet the requirements of the
        chart.")

}
```

The validation test performs two tests: the *noPlayerDups* test and the *columnTest.* If both tests evaluate to *TRUE*, then the code that creates the chart is executed. Otherwise, a blank chart is created with information about why *ggplot2* was not able to create the chart.

Step 9: Create the chart titles

Static strings are used in this example for the chart title, chart subtitle, and caption. They are defined using the following code:

```
chart.title <- "LA Lakers player comparison"
chartsubtitle <- "(2008 - 2009 Season)"
chart_source <- "Source:  lubridate package"
```

Step 10: Add additional columns to the data set

You are creating a quad chart that compares the LA Lakers basketball players based on their rebounds and points scored. Rebounds and points scored are on two different scales because the range of possible values for the number of points scored is a lot higher than the range of possible values for the number of rebounds. The quad chart requires them to be on the same scale. That is done in this step using the following code:

```
graph_data <-
  dataset %>%
  mutate(
    Scaled.Rebounds =
      round(rescale(Rebounds, to = c(-10, 10)), 1)
    ,Scaled.TotalPoints =
      round(rescale(`Total Points`, to = c(-10, 10)), 1)
  )
```

The *mutate* verb is used in this step to add two new columns to the data frame: *Scaled.Rebounds* and *Scaled.TotalPoints*. Those two columns use the *rescale()* function from the *scales* package to put them both on a scale between –10 and +10. These new columns will be used in the visualization to represent the rebounds and total points scored for each player.

Step 11: Start the chart by defining the ggplot function

The following code is used to define the *ggplot()* function:

```
p <- ggplot(graph_data,
          aes(x = Scaled.Rebounds, y = Scaled.TotalPoints)
  )
```

The data set that will be used in this visual is the *graph_data* data frame and only two aesthetic are defined in the *ggplot()* function. They are the *x* and *y* aesthetics.

Step 12: Use the geom_point() geom to plot the players on the plot

Next, you add dots to the graph to represent the players. The positions of the dots are based on the rescaled version of the number of rebounds and the total number of points the given player has. This is done by adding a *geom_point()* geom to the code as depicted here:

```
p <- ggplot(
            graph_data,
            aes(x = Scaled.Rebounds, y = Scaled.TotalPoints)
      ) +
      geom_point()
```

The *geom_point()* geom inherits the aesthetics from the *ggplot()* function so the required *x* and *y* aesthetics do not need to be defined in the *geom_point()* layer.

Step 13: Add the labels for each quadrant

Next, you add a layer of labels to the visual that gives the name of the players the dots represent. This is done using the *geom_label_repel()* geom as depicted in the following code:

```
ggplot(graph_data,
      aes(x = Scaled.Rebounds,
          y = Scaled.TotalPoints)
      ) +
      geom_point() +
      geom_label_repel(aes(
          label = Player),
          size = 4, show.legend = FALSE
      )
```

The *geom_label_repel()* geom inherits the *x* and *y* argument from the *ggplot()* function so you do not need to define those two aesthetics. The only aesthetic needed to be defined is the *label* aesthetic, which you set to the *Player* column. The values in this column will be used to label the dots with the name of the player they represent. You set the *size* argument outside the *aes()* function because you want all the labels to have the same font size. Also, by default, *ggplot2* creates a legend for each scale added outside the *x* and *y* scale. A legend is not needed for the *label*, so to prevent one from being shown, you set the *show.legend* property equal to *FALSE*.

Step 14: Draw vertical and horizontal lines through the x and y axis

The visual requires distinct lines for the *x* and *y* axis. The last two lines in the following code add the required horizontal and vertical lines:

```
p <- ggplot(
            graph_data,
            aes(x = Scaled.Rebounds,
                y = Scaled.TotalPoints)
               ) +
      geom_point() +
      geom_label_repel(
          aes(label = Player),
          size = 4,
          show.legend = FALSE
      ) +
      geom_hline(yintercept = 0) +
      geom_vline(xintercept = 0)
```

The *geom_hline()* and *geom_vline()* geoms are used to draw the lines. Setting the *yintercept* in the *geom_hline()* function to 0 draws a horizontal line that goes through point 0 of the y axis, and setting *xintercept* in the *geom_vline()* function to 0 draws a vertical line that goes through point 0 of the *x* axis.

Step 15: Add quadrant labels to the chart

Figure 2-32 shows what the visual looks like after performing the previous steps.

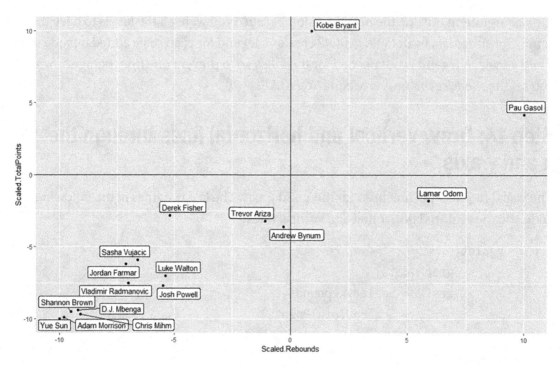

Figure 2-32. *The beginnings of the quad chart*

The chart is coming along great. You are making good progress toward your end goal. In this step, you will add labels to each quadrant using the following code:

```
annotate("text", x = -5, y = 11, label = "High Scorer",
        alpha = 0.2, size = 6) +
annotate("rect", xmin = -3.5, xmax = -6.5, ymin = 10.5,
        ymax = 11.5, alpha = .2) +

annotate("text", x = 5, y = 11, label = "Beast Mode",
        alpha = 0.2, size = 6) +
annotate("rect", xmin = 3.5, xmax = 6.5, ymin = 10.5,
        ymax = 11.5, alpha = .2)

annotate("text", x = -5, y = -11, label = "Average",
        alpha = 0.2, size = 6) +
annotate("rect", xmin = -3.5, xmax = -6.5, ymin = -11.5,
        ymax = -10.5, alpha = .2) +
```

```
annotate("text", x = 5, y = -11, label = "High Rebounder",
        alpha = 0.2, size = 6) +
annotate("rect", xmin = 3.5, xmax = 6.5, ymin = -11.5,
        ymax = -10.5, alpha = .2) +
```

The preceding code builds labels with borders for each quad in the quad chart. The first quad in the upper left is labeled *High Scorer* because players in this group rank high in scoring but low in rebounding, the second quad in the upper right is named *Beast Mode* because players in this quadrant rank high in both rebounding and scoring, the third quadrant in the lower left is labeled *Average* because players in this quadrant rank relatively low in both rebounding and scoring, and the fourth quadrant in the lower right is labeled *High Rebounder* because players in this quadrant rank high in rebounds but low in scoring.

Both the text and the border of the quadrant labels are created using the *annotate()* function. The first argument of the *annotate()* function is where you specify the type of geom you want to use for your annotation. The types of geoms that are being used in this example to create the quadrant labels are the *text* and *rect* geoms. In the preceding code, the text of the label for the respective quadrant is created using the *text* geom, then a rectangular border is placed around the text using the *rect* geom. First, let's go over how the *text annotation* works, then we will go over how the *rect annotation* works.

As stated earlier, you create a *text annotation* by setting the value of the first argument in the *annotate()* function to *text*. The *x* and *y* argument are used to determine the position of the "text". In this visual, the *x* values of the *text annotations* are positioned at the center of their respective quadrants, and the *y* values of the *text annotations* are positioned just outside of the area of possible data points. The latter is done to minimize the player data from overlapping with the quadrant labels. The *alpha* argument of each text annotation is set to 0.2 to make sure the player names would show over the labels if they were in the same position.

The *alpha* argument is used to set the transparency of a geom on a chart. The closer to zero, the more transparent the element will be, and the closer to one, the more opaque the element will be.

After the quadrant label is created using the *text annotation* for the given quadrant, the border of the label is created using the *rect annotation*. This is accomplished by setting the first argument of the *annotate()* function to "rect". The rectangle that you want to draw is defined by using the *xmin, xmax, ymin,* and *ymax* arguments. Here are descriptions that explain the purpose of each argument:

- The *xmin* argument is used to determine the *x* position of the upper left and lower left corners.

- The *xmax* argument is used to determine the *x* position of the upper right and lower right corners.

- The *ymin* argument is used to determine the *y* position of the lower left and lower right corners.

- The *ymax* argument is used to determine the *y* position of the upper left and upper right corners.

Once the corners are defined, the *annotate()* function connects the points with lines. The result is a rectangular border around the labels.

Step 16: Add labels for the x and y axis

Here is the code used to add the arrows to the x and y axis label names:

```
xlab(bquote("Rebounding" ~ symbol('\256'))) +
ylab(bquote("Scoring" ~ symbol('\256'))) +
```

This is a hack in which I used the *bquote()* function that is typically used to create mathematical expressions in R. In this example, you are using it not to create a mathematical expression but to add the arrow symbol to the labels. This is accomplished via the *xlab()* and *ylab()* functions. The *xlab()* function is used to rename the x axis label, and the *ylab()* function is used to rename the y axis label. The code symbol('\256') is the code needed to draw an arrow.

Step 17: Add the dynamic titles and caption to the custom R visual

The chart title, the subtitle, and the chart source are added to the visual using the *labs()* function using the following code:

```
labs(
    title = chart.title,
    subtitle = chartsubtitle,
    caption = chart_source
)
```

Step 18: Add a theme

The theme of the chart is changed using visualization principles from Edward Tufte via the *theme_tufte()* theme from the *ggthemes* package. This is accomplished by adding the following code to the script:

```
theme_tufte() +
```

Step 19: Perform last minute cleanup

Performing the previous steps will result in the visualization in Figure 2-33.

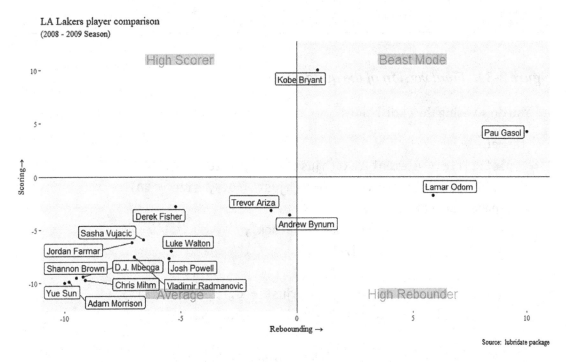

Figure 2-33. *Quad chart prior to last minute cleanup*

You want to reformat the chart titles and the labels of the axes to look like the visual below.

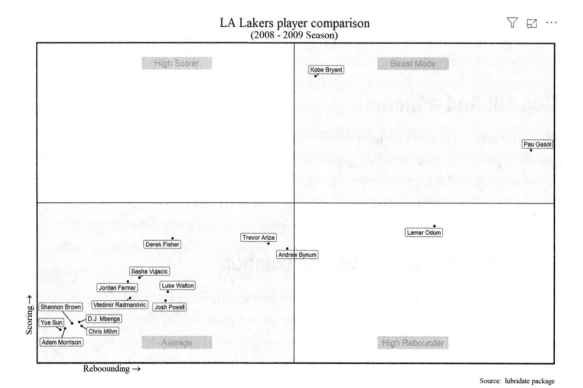

Figure 2-34. *Final version of the quad chart*

You do so using the code below:

```
theme(
  plot.title = element_text(hjust = 0.5, size = 25)
  , plot.subtitle = element_text(hjust = 0.5, size = 20)
  , panel.border = element_rect(
                      colour = "black",
                      size = 2,
                    fill = NA)
  , axis.title.x = element_text(hjust = 0.1, size = 18)
  , axis.title.y = element_text(hjust = 0.1, size = 18)
) +
scale_x_continuous(labels = NULL, breaks = NULL) +
scale_y_continuous(labels = NULL, breaks = NULL)
```

Let's go over the preceding code in detail. As you can see, most of the required changes are handled using the *theme()* function. Here is a list of the components that are modified by the *theme()* function in this step along with a description of how they were modified:

- *plot.title*: This component is used to control the chart title, and it uses the *element_text()* function to do so. The *hjust* argument of the *element_text()* function is used to control the horizontal position of the text, and it accepts values between 0 and 1. A value of 0 justifies the text to the left, a value of 0.5 centers the text, and a value of 1 puts the value to the far right. You use a value of 0.5 because you want the text to be centered. You use the *size* argument to increase the size of the font to 25.

- *plot.subtitle*: This component is used to control the chart subtitle, and it also uses the *element_text()* function to do so. The *hjust* argument is set the same way as it was in the *plot.title* component because you want to center the text as you did in the *plot.title* component. You want the subtitle to be smaller than the main title so you set it to the lower number of 20.

- *panel.border*: This component is used to control the border around the visual, and it uses the *element_rect()* function to do so. The *colour* argument is used to set the color of the border, and the *size* argument is used to set the thickness of the border.

- *axis.title.x*: This component is used to control the label of the x axis, and it uses the *element_text()* function to do so. You want to label it to be positioned on the left side of the x axis so you use a value of 0.1 for the *hjust* argument to do so. You use the *size* argument to change the font size of the text to 18.

- *axis.title.y*: This component is used to control the label of the y axis, and it is configured similarly as the axis.title.x. The difference is that the positioning is on a vertical line instead of a horizontal line. Because of that, the closer you are to zero the lower your position will be on the Y axes and the closer you are to 1 the higher your position will be on the Y axis. You are using a value of 0.1 so that positions your Y axes label close to the origin.

Lastly, the *scales_x_continuous()* and *scale_y_continuous()* scales are used to reformat the axes. The labels and breaks in both scale functions are set to NULL to remove them from the chart. The results of these actions were illustrated earlier in Figure 2-34.

Step 20: Add code to Power BI

The complete script to produce the visual is in Listing 2-7.

Listing 2-7. The R script that produces the quad chart

```
library(tidyverse)
library(ggrepel)
library(ggthemes)
library(scales)

currentColumns <- sort(colnames(dataset))
requiredColumns <- c("Player", "Rebounds", "Total Points")
columnTest <- isTRUE(all.equal(currentColumns, requiredColumns))
noPlayerDups <- length(unique(dataset$Player)) == length(dataset$Player)

if (noPlayerDups & columnTest) {

  chart.title <- "LA Lakers player comparison"
  chartsubtitle <- "(2008 - 2009 Season)"
  chart_source <- "Source:  lubridate package"

  graph_data <-
    dataset %>%
    mutate(
      Scaled.Rebounds =
        round(rescale(Rebounds, to = c(-10, 10)), 1)
      ,Scaled.TotalPoints =
        round(rescale(`Total Points`, to = c(-10, 10)), 1)
    )

  p <- ggplot(graph_data, aes(x = Scaled.Rebounds, y = Scaled.TotalPoints)) +
    geom_point() +
    geom_label_repel(aes(label = Player), size = 4, show.legend = FALSE) +
```

```
geom_hline(yintercept = 0) +
geom_vline(xintercept = 0) +

# quad 1
annotate("text", x = -5, y = 11, label = "High Scorer", alpha = 0.2,
size = 6) +
annotate("rect", xmin = -3.5, xmax = -6.5, ymin = 10.5, ymax = 11.5,
alpha = .2) +

# quad 2
annotate("text", x = 5, y = 11, label = "Beast Mode", alpha = 0.2,
size = 6) +
annotate("rect", xmin = 3.5, xmax = 6.5, ymin = 10.5, ymax = 11.5,
alpha = .2) +

# quad 3
annotate("text", x = -5, y = -11, label = "Average", alpha = 0.2,
size = 6) +
annotate("rect", xmin = -3.5, xmax = -6.5, ymin = -11.5, ymax = -10.5,
alpha = .2) +

# quad 4
annotate("text", x = 5, y = -11, label = "High Rebounder", alpha = 0.2,
size = 6) +
annotate("rect", xmin = 3.5, xmax = 6.5, ymin = -11.5, ymax = -10.5,
alpha = .2) +

# Shade lower left quadrant
annotate("rect", xmin = -Inf, xmax = 0.0, ymin = -Inf, ymax = 0,
alpha = 0.1, fill = "lightskyblue") +

# Shade upper right quadrant
annotate("rect", xmin = 0.0, xmax = Inf, ymin = 0.0, ymax = Inf,
alpha = 0.1, fill = "lightskyblue") +

# Titles
xlab(bquote("Rebounding" ~ symbol('\256'))) +
ylab(bquote("Scoring" ~ symbol('\256'))) +
```

```
    labs(title = chart.title, subtitle = chartsubtitle, caption = chart_
    source) +

    # Prettying things up
    theme_tufte() +
    theme(
      plot.title = element_text(hjust = 0.5, size = 25)
      , plot.subtitle = element_text(hjust = 0.5, size = 20)
      , panel.border = element_rect(colour = "black", size = 2, fill = NA)
      , axis.title.x = element_text(hjust = 0.1, size = 18)
      , axis.title.y = element_text(hjust = 0.1, size = 18)
    ) +
    scale_x_continuous(labels = NULL, breaks = NULL) +
    scale_y_continuous(labels = NULL, breaks = NULL)

  p
} else {

  plot.new()
  title("The data supplied did not meet the requirements of the chart.")

}
```

Copy and then paste the script in the *R script editor* associated with this visual. The resulting R visual is totally responsive to the filters in the report.

Adding regression line

One of the most common enhancements that data analysts tend to add to their scatter plots are *trend lines*. Adding a *trend line* to your scatter plot in R using *ggplot2* is straightforward, and you have more configuration options than you typically have in other visualization tools. For instance, you have the option to

- Use *linear models, generalized linear models, generalized additive models,* and a few others for your trend line

- Determine what you want your confidence level to be

- Use your own formula for your trend line

- Use many more configurable options that are too numerous to list

In this section, you will use a simple example based on a data set that comes pre-installed in R. The name of the data set is *women,* and it is a data set that shows the average weight by height for women in the United States. You will plot the information using a scatter plot, then overlay the scatter plot with a trend line based on a linear regression model. The resulting visual is listed in Figure 2-35.

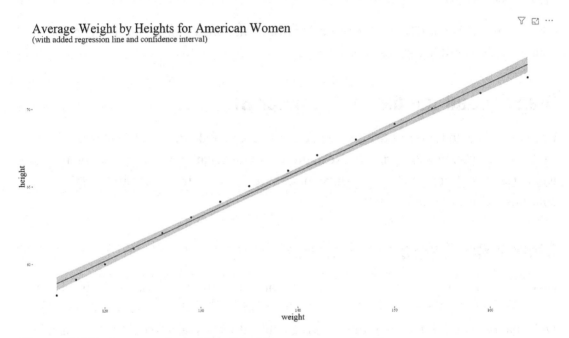

Figure 2-35. *Scatter plot with added linear regression line*

Let's go over the steps to produce this visual!

Step 1: Acquire the necessary data

The code used in this example is one of the data sets that come pre-installed with base R. Here is the R code used to create the *csv* file that is the source data for this example:

```
library(tidyverse)

data(women)
setwd("<path to folder where the file is located>")
write_csv(women, "women.csv")
```

A file named *women.csv* is the result of running the preceding code, and it will be the source for Power BI in the next step.

Step 2: Load the data into Power BI

Read the data into Power BI using *GetData*. To do so, click the *Home* tab in Power BI, then click *GetData* ➤ *Text/CSV*. Next, browse to the location where you saved the *women.csv* file. When you load the data, make sure that both the *height* and *weight* columns are in a number format.

Step 3: Configure the R visual

Go to the visualization pane and drag an R visual to the report canvas. Resize the visual to the desired size. Drop the *height* and *weight* columns from the *women* table into the *Fields* pane. If the *Fields* pane is not showing, select the R visual and it should appear.

Step 4: Export data to R Studio for development

Click the 45° arrow that is located on the title bar of the *R script editor*. This action will export the data frame passed to R and the R starter code to *R Studio*. Please refer to Step 4 in the "Callout chart" section if you need a more detailed explanation.

Step 5: Load the required packages

The R packages that will be used in this script are *tidyverse* and *ggthemes*. The *ggplot* package from *tidyverse* is used to create the visual, and the *dplyr* package from *tidyverse* is used to shape the data for the visual. The *ggthemes* package provides the *theme_tufte()* theme that will be used to change the chart's theme.

Step 6: Create the variables needed for the data validation test

The following code is used to create the variables for the data validation test:

```
currentColumns <- sort(colnames(dataset))
requiredColumns <- c("height", "weight")
columnTest <- isTRUE(all.equal(currentColumns, requiredColumns))
```

You need to make sure the data set passed to R from Power BI has the proper structure, and you need to be prepared to handle the situation if it is not. The test you perform in this example is a test that makes sure that the data set passed to R contains only the columns needed for the R visual.

You do so by first getting the names of the columns in the *dataset* data frame, and you assign it to a character vector variable named *currentColumns*. The elements in the variable are sorted in alphabetical order. The next line of code creates another character vector that contains the columns that are needed to create the R visual. They are also supplied in alphabetical order. The last line in the code snippet performs a test to see if the *currentColumns* and *requiredColumns* character vectors are equal using the *all.equal()* function. The *all.equal()* function is used to compare two R objects to see if they are equal. If they are equal, then TRUE is returned; otherwise, information is returned that attempts to explain why they are unequal. In this example, you are just concerned if the *currentColumns* and *requiredColumns* variables are equal. You just need a Boolean response. You can get that by wrapping *all.equal()* with the *isTRUE()* function.

Step 7: Create the data validation test

Here is the shell of the data validation test:

```
if (columnTest) {
        <code to produce visual>
} else {
  plot.new()
  title("The data supplied did not meet the requirements of the
        chart.")
}
```

Recall the *columnTest* variable is based on a test that returns a Boolean value. If the value of the *columnTest* variable is TRUE, then the code will be executed to build the visual; otherwise, the code in the *else* block will be executed. The code in the *else* block generates a blank plot using the *plot.new()* function and adds text to the blank plot that explains to the report user why the desired chart was not created.

Step 8: Start the chart by initializing the ggplot function

Here is the code used in this step:

```
chartdata <- dataset

ggplot(chartdata, aes(x = weight, y = height)) +
```

First, you make a copy of the *dataset* data frame by assigning it to a variable named *chartdata*. Next, you define the *ggplot()* function. The first argument is the data set that the visual will use. The second argument uses the *aes()* function to define a couple visual aesthetics. The aesthetics that are being defined are *x* and *y* coordinates.

Step 9: Add a scatter plot layer to the R visual

The scatter plot layer is added by using the *geom_point()* geom. The values for the *x* and *y* arguments are inherited from the *ggplot()* function and are used to determine the position of each point in the scatter plot. Here is the code used to add the geom:

```
geom_point() +
```

The result of adding the *geom_point()* geom to the chart is illustrated in Figure 2-36.

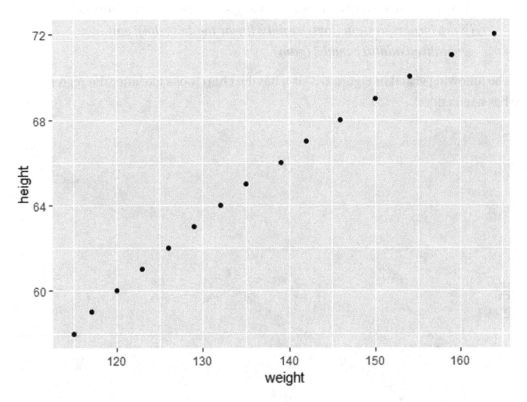

Figure 2-36. *Scatter plot of the average American women weight by height using the "women" data set that comes with base R*

Step 10: Add regression line layer to the R visuals

In this step, you add a regression line layer using the *geom_smooth()* geom. The code used to add the layer is as follows:

```
geom_smooth(method='lm') +
```

The preceding code will add a linear regression line to the chart based on the data in the scatter plot. What makes this function so special is that it is very configurable. The *geom_smooth()* function enables you to

- Specify your own linear regression formula

- Add a confidence interval to your linear regression line

- Choose the confidence interval level you want to use

- Use other smoothing methods such as *generalized linear regression (glm), local regression (loess), robust linear models (rlm),* and *generalized additive model (gam)*

The following visual in Figure 2-37 is what the chart looks like after the regression line has been added.

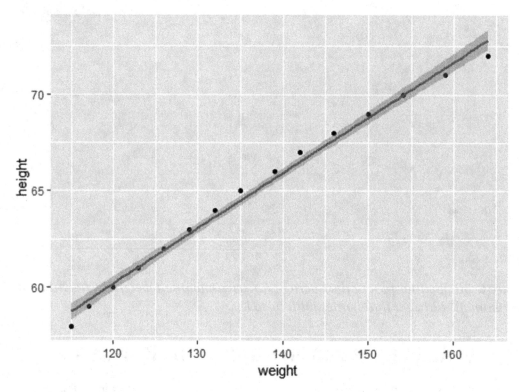

Figure 2-37. *Scatter plot with added regression line*

Step 11: Add a title to the chart

The *ggtitle()* function is used to add the chart title and subtitle as shown here:

```
ggtitle(
    "Average Weight by Heights for American Women",
    subtitle = "(with added regression line and
                confidence interval)"
) +
```

Note the *ggtitle()* function is an alternative to the *labs()* function, but it has the required functionality needed in this example.

Step 12: Change the chart's theme

You want to change the look and feel of the chart using principles from Stephen Few. You do so using the following code which adds the *theme_tufte()* theme to the chart:

```
theme_tufte() +
```

Step 13: Perform some last minute cleanup

The *theme()* function is used to change the font size of the chart title, subtitle, and axis labels. The chart title's font is changed using the *plot.title* component, the chart's subtitle is changed using the *plot.subtitle* component, and the axis labels are changed using the *axis.title* component. All three components use the *element_text()* function to make the changes:

```
theme(
  axis.title = element_text(size = 20),
  plot.title = element_text(size = 30),
  plot.subtitle = element_text(size = 20)
)
```

Step 14: Add code to Power BI

The complete script is shown in Listing 2-8. Copy the script and paste it in the *R script editor* in Power BI.

Listing 2-8. The R script that produces the scatter chart with regression line

```
library(tidyverse)
library(ggthemes)

currentColumns <- sort(colnames(dataset))
requiredColumns <- c("height","weight")
columnTest <- isTRUE(all.equal(currentColumns, requiredColumns))
```

```r
if (columnTest) {

  chartdata <- dataset

  ggplot(chartdata, aes(x = weight, y = height)) +
    geom_point() +
    geom_smooth(method='lm') +
    ggtitle(
      "Average Weight by Heights for American Women",
      subtitle = "(with added regression line and confidence
                  interval)"
    ) +
    theme_tufte() +
    theme(
      axis.title = element_text(size = 20),
      plot.title = element_text(size = 30),
      plot.subtitle = element_text(size = 20)
    )

} else {

  plot.new()
  title("The data supplied did not meet the requirements of the chart.")

}
```

In this chapter, you walked through several recipes that illustrated how expressive you can be with your visualizations in Power BI via the *ggplot2* package. You can easily modify the recipes to fit your needs, or you can use them to inspire a completely different visual. The *ggplot2* package along with its helper packages provides a great framework to build R visuals for Power BI in a relatively low code way. The *ggplot2* package is a perfect companion to Power BI!

PART II

Ingesting Data into the Power BI Data Model Using R and Python

Reading CSV Files

Power BI is an excellent tool for data visualization and has been recognized as one of the premier tools in that category. However, before you are able to create the great visualizations that Power BI is known for, you first need to get data into the Power BI data model.

Power BI comes with a tool named Power Query that enables you to handle most situations you will encounter while bringing data into the Power BI data model. Power Query's graphical user interface is intuitive and easy to use, but there are advanced situations where Power Query will fall short.

Fortunately, with the addition of R and Python in Power BI, many of those "hard-to-do" tasks become relatively easy to handle. In this chapter, you will learn how to use R and Python to dynamically choose the files in a folder that you want to combine and filter rows based on a string pattern that requires advanced string matching techniques. Let's start with dynamically combining files.

Dynamically combining files

Combining multiple files from a folder into a data set prior to bringing it into the Power BI data model is a common task, but choosing those files dynamically using complex selection logic can be challenging in Power Query. This section shows how to make the task easy through using R and Python.

Combining files is straightforward in Power Query if you are combining all the files in a particular folder into one data set when the files share the same data structure. Combining files is also not a difficult task if you are combining files in a folder into one data set using a simple selection rule to pick the files you want to combine. But when you need to apply a selection rule that involves complex logic, then the task may be hard if not impossible to do in Power Query. Using R and Python makes the task easier to perform, even when the selection criteria is complex.

© Ryan Wade 2020
R. Wade, *Advanced Analytics in Power BI with R and Python*, https://doi.org/10.1007/978-1-4842-5829-3_3

Example scenario

Here is the scenario. It is February 5, 2014, and you are a financial analyst for a major widget seller. You have a folder that contains monthly sale reports dating from 2010-12-01 to 2014-01-01. Each month's report is relatively big, and because of limitations of Power BI, you are not able to load all historical reports. You need to limit your load to a rolling 24-month period.

Here are some additional technical details:

- The reports are in a *csv* file type.

- The naming convention used for the file names is *YYYY-MM-DD.csv.*

- All the reports have the same structural layout.

Creating a workflow in Power Query that can dynamically pick files that are in the rolling 24-month period would be hard to do. The task requires advanced Power Query skills and the use of custom M code. The underlying M code needed to perform this task would be complicated and not as intuitive as the code would be if R or Python were used. Let's first perform this task using R, then we will perform the same task using Python. That way, you can see how intuitive and concise the code is using both of those languages.

Picking the rolling 24 months using R

R users have been doing *data wrangling* since the early 1990s. Data wrangling is the process of taking raw data and transforming it into a data set that is conducive for analysis. The process of combining like files into one data set is a common data wrangling task. The R community has developed many packages over the years that makes such tasks much easier to do. You will see it being illustrated in this example. Let's go over the steps needed to perform the stated task in R.

Step 1: Load the required R packages for the script

In this example, you will use four packages from *tidyverse*. Those packages are *readr*, *lubridate*, *purrr*, and *stringr*. The *readr* package provides functionality that facilitates reading flat files into R, the *lubridate* package provides functions that make it easier to work with dates in R, the *purrr* package implements functional programming techniques in R in a consistent way, and the *stringr* package provides some handy string manipulation functions.

The *readr* and *purrr* packages are included in the *tidyverse* load so you do not need to explicitly load them. You do, however, need to load *lubridate* and *stringr* because they are not part of the core packages that are included in the *tidyverse* load. Here is the code needed to load the packages:

```
library(tidyverse)
library(lubridate)
library(stringr)
```

You may be thinking that, if you need to use four R packages for one script, you are actually making the situation harder because you don't need to load additional packages in Power Query. Everything you need in Power Query is already built in. Please follow along with me and you will see that having access to a large number of packages in R is a big plus. You will be able to perform some complicated tasks with very little code by leveraging R packages.

Step 2: Change your working directory to the folder that contains the Sales data sets

A *working directory* in R is the root directory where your scripts and data are located. Using a working directory is recommended because it enables you to keep all items that are needed for your task in one central location. It also enables you to take advantage of *relative file path referencing*. *Relative file path referencing* enables you to reference a file relative to your working directory without having to supply the full file path. Your working directory for this example should be set to the folder named *R_Code* which is a subfolder in the *Chapter03* section of the repo. The *R_Code* folder is contained in the code download for this book. So if you saved the companion code to your *Documents* folder, then you would set the working directory with code that looks similar to the following code:

```
setwd("C:/Users/<"username">/Documents/AdvancedAnalyticsPowerBI/
Chapter03/R_Code")
```

The *setwd()* function sets your working directory to a new location based on the file path that you passed to it.

If you want to set the working directory to a location relative to where your current working directory is, you do so by starting your file path string with a period (". "). Let's illustrate with an example. The code that follows assumes that your working directory is your *Documents* folder and you want to change it to the *R_Code* folder using forward slashes:

```
setwd("./AdvancedAnalyticsPowerBI/R_Code ")
```

The preceding code tells R to start from the *Documents* folder, your working directory, via the use of the period. From there, you can traverse to the folder you want to set your new working directory to using "/", followed by the folder's name. You can also use "\", but if you do, you need to use two backslashes because "\" is a special character that needs to be *escaped*.

When you literally want to use a character that is also a special character, you need to tell R and Python that you want to use the literal version of it by escaping it with a "\". Some commonly used special characters that need to be escaped are "*" and ".".

Step 3: Read the file names into a character vector

If you want to select the files that should be included in the rolling 24 months, you need to be able to analyze the file names to determine whether they should be included. To do that, you use the *list.files()* function. The *list.files()* function gets all the file names in the file path that is passed to it and puts it into a *character vector* as illustrated in the following code:

```
monthly_reports <- list.files("./Data/SalesData/")
```

A *character vector* is a vector of data whose elements are of the *character* data type. The *character* data type in R is similar to the *string* data type in DAX and the *text* data type in M.

In the preceding code, R first gets all the file names in the specified directory and puts them into a *character vector*. It then assigns the resulting character vector to a variable named *monthly_reports*.

The <- is an assignment operator used to assign values to R variables.

Here is a printout of the information that is stored in the *monthly_reports* variable:

```
> monthly_reports
 [1] "2010-12-01.csv" "2011-01-01.csv" "2011-02-01.csv"
 [4] "2011-03-01.csv" "2011-04-01.csv" "2011-05-01.csv"
 [7] "2011-06-01.csv" "2011-07-01.csv" "2011-08-01.csv"
[10] "2011-09-01.csv" "2011-10-01.csv" "2011-11-01.csv"
[13] "2011-12-01.csv" "2012-01-01.csv" "2012-02-01.csv"
[16] "2012-03-01.csv" "2012-04-01.csv" "2012-05-01.csv"
[19] "2012-06-01.csv" "2012-07-01.csv" "2012-08-01.csv"
[22] "2012-09-01.csv" "2012-10-01.csv" "2012-11-01.csv"
[25] "2012-12-01.csv" "2013-01-01.csv" "2013-02-01.csv"
[28] "2013-03-01.csv" "2013-04-01.csv" "2013-05-01.csv"
[31] "2013-06-01.csv" "2013-07-01.csv" "2013-08-01.csv"
[34] "2013-09-01.csv" "2013-10-01.csv" "2013-11-01.csv"
[37] "2013-12-01.csv" "2014-01-01.csv"
```

Step 4: Create a date vector

Now create a vector with the date representation of each file name. In order to be able to determine the rolling 24 months, you need to get the file names into a date format. You accomplish this in two steps. In the first step, you remove ".csv" so that you are left with a string that can be parsed into a date. To do that, you use the *str_replace()* function from the *stringr* package and assign the output to a character vector named *date_format*. Here is the code:

```
date_format <- stringr::str_replace(monthly_reports, ".csv","")
```

The *str_replace()* function requires three parameters: the string you want to change, the pattern in the string you want to replace, and the string you want to use as a replacement. Here is a printout of what is held in the date_format variable created earlier:

```
> date_format
  [1] "2010-12-01" "2011-01-01" "2011-02-01" "2011-03-01"
  [5] "2011-04-01" "2011-05-01" "2011-06-01" "2011-07-01"
  [9] "2011-08-01" "2011-09-01" "2011-10-01" "2011-11-01"
 [13] "2011-12-01" "2012-01-01" "2012-02-01" "2012-03-01"
 [17] "2012-04-01" "2012-05-01" "2012-06-01" "2012-07-01"
 [21] "2012-08-01" "2012-09-01" "2012-10-01" "2012-11-01"
 [25] "2012-12-01" "2013-01-01" "2013-02-01" "2013-03-01"
 [29] "2013-04-01" "2013-05-01" "2013-06-01" "2013-07-01"
 [33] "2013-08-01" "2013-09-01" "2013-10-01" "2013-11-01"
 [37] "2013-12-01" "2014-01-01"
```

Note that *stringr::* before the *str_replace()* function is not needed. It is there to make it clear which package the function came from.

The next thing you need to do is parse the string into a date. You can do that by using the *ymd()* function from the *lubridate* package. Here is the code needed to do so:

```
date_format <- lubridate::ymd(date_format)
```

The *ymd()* function is a pretty slick function! It parses a string that is in a year-month-day format to a date, and it is very forgiving. It does a good job of parsing dates that come from date string that are harder to parse by other programming languages such as M in Power Query. You take the resulting date vector and assign it to the *date_format* variable. Here is a display of the contents held in the *date_format* variable:

```
> date_format
  [1] "2010-12-01" "2011-01-01" "2011-02-01" "2011-03-01"
  [5] "2011-04-01" "2011-05-01" "2011-06-01" "2011-07-01"
  [9] "2011-08-01" "2011-09-01" "2011-10-01" "2011-11-01"
 [13] "2011-12-01" "2012-01-01" "2012-02-01" "2012-03-01"
 [17] "2012-04-01" "2012-05-01" "2012-06-01" "2012-07-01"
```

```
[21] "2012-08-01" "2012-09-01" "2012-10-01" "2012-11-01"
[25] "2012-12-01" "2013-01-01" "2013-02-01" "2013-03-01"
[29] "2013-04-01" "2013-05-01" "2013-06-01" "2013-07-01"
[33] "2013-08-01" "2013-09-01" "2013-10-01" "2013-11-01"
[37] "2013-12-01" "2014-01-01"
```

The output appears to be the same as it was before the *ymd()* function was applied, but it is actually not. You can prove that using the *str()* function to inspect it. The *str()* function is used to display the structure of an R object. If you pass the *date_format* variable to the *str()* function, you will get the following result which proves it is a *date* data type:

```
> str(date_format)
 Date[1:38], format: "2010-12-01" "2011-01-01" "2011-02-01"
```

It is recommended that you perform your testing in the *R console* and not in the R script *editor* in R Studio to prevent adding code to your script that should not be there.

Step 5: Create a data frame using the two vectors

You want to put the information into data frame because having the data in a data frame makes analysis much easier. An R data frame is a two-dimensional tabular object made up of variables (columns) that can have different data types. They are similar to *Microsoft Excel* tables and SQL Server tables in some ways, but they are specifically designed for data analysis. The code needed to convert the *monthly_reports* variable and *date_format* variable to a data frame is listed as follows:

```
df <- tibbles::data_frame(monthly_reports, date_format)
```

You use *data_frame()* function from the *tibble* package instead of *data.frame()* function that is part of base R because *data.frame()* has some undesirable side effects. The *data_frame()* function creates a type of object known as a *tibble*, which can be thought of as the modern version of the data frame. I find tibbles to be more desirable to work with.

One of the attributes that I like about *tibbles* is the way they print to the screen. When you print a *tibble*, only the first ten rows are outputted, and only the columns that can fit on the screen are shown. When you print a regular data frame, R displays a lot more information and that can be a problem if you are dealing with a big data set. You also get the data type of each column from a tibble, which is valuable information. Here is what the output looks like when I print the *df* data frame:

```
> df
# A tibble: 38 x 2
   monthly_reports date_format
   <chr>           <date>
 1 2010-12-01.csv  2010-12-01
 2 2011-01-01.csv  2011-01-01
 3 2011-02-01.csv  2011-02-01
 4 2011-03-01.csv  2011-03-01
 5 2011-04-01.csv  2011-04-01
 6 2011-05-01.csv  2011-05-01
 7 2011-06-01.csv  2011-06-01
 8 2011-07-01.csv  2011-07-01
 9 2011-08-01.csv  2011-08-01
10 2011-09-01.csv  2011-09-01
# ... with 28 more rows
```

Step 6: Get the upper and lower bound of our desired date range

You want a rolling 24-month window. Thus, you need to compute the upper and lower bound of the date range needed to define that window. You can do so using the following code:

```
max_month <- max(df$date_format)
min_month <- max_month %m+% lubridate::months(-23)
```

You use the *max()* function over the *date_format* column in the *df* data frame to get the max month, and you assign the results to the *max_month* variable. Next, you take the *max_month* variable and subtract 23 months from it to get the min month by invoking the *%m+%* operator from the *lubridate* package. That operator is designed to add months. In this example, you are adding –23 months. The resulting value is assigned to the *min_month* variable.

Step 7: Subset the data frame to only include the desired months

In this step, you apply a filter that returns the vector that contains the files you want to combine. You do so by applying a filter that returns only files for the months that are greater than or equal to your minimum month. Here is the code:

```
reports_to_read <-
  df$monthly_reports[df$date_format >= min_month]
```

In R, you are able to filter a column in a data frame using the preceding method. The code in between the brackets, df$date_format >= min_month, returns an R object called a *Boolean* vector. In the expression, each element in the *date_format* column is compared to the *min_month* variable and *True* is returned in the situations where the element in the column is greater than or equal to *min_month*, and *False* is returned otherwise. Only the rows that evaluate to *True* are kept.

You only want the *reports_to_read* column. You subset that column in the *df* data frame by using df$monthly_reports syntax. If you wanted to return all the columns in the data frame, you could have done so with the following code:

```
reports_to_read <- df[df$date_format >= min_month]
```

This is a method for subsetting data that relies on base R functionality, and it is fine in simple situations. You will learn more intuitive methods later in the book that can be used for more complicated subsetting scenarios. The contents of the *reports_to_read* variable are shown here:

```
> reports_to_read
 [1] "2013-01-01.csv" "2013-02-01.csv" "2013-03-01.csv"
 [4] "2013-04-01.csv" "2013-05-01.csv" "2013-06-01.csv"
 [7] "2013-07-01.csv" "2013-08-01.csv" "2013-09-01.csv"
[10] "2013-10-01.csv" "2013-11-01.csv" "2013-12-01.csv"
[13] "2014-01-01.csv"
```

Step 8: Create a data frame that is based on the union of all the files

You now have the list of files that you want to combine into one data frame in the *reports_to_read* variable. Now it's time to actually do the combining. You can easily combine the files with one line of code using the *map_df()* function from the *purrr* package in R. Here is that line of code:

```
df_output <- purrr::map_df(reports_to_read, read_csv)
```

This line of code uses the *map_df()* function to map each element from the *reports_to_read* variable to the *read_csv()* function from the *readr* package. The *_df* portion of the function name tells you that the result from the function will be an R data frame that contains the data from all of the files. Here is what the output of the resulting data frame looks like:

```
> df_output
# A tibble: 54,771 x 9
   Category OrderDate          DueDate
   <chr>    <dttm>             <dttm>
 1 Bikes    2013-01-01 05:00:00 2013-01-13 05:00:00
 2 Accesso~ 2013-01-01 05:00:00 2013-01-13 05:00:00
 3 Accesso~ 2013-01-01 05:00:00 2013-01-13 05:00:00
 4 Bikes    2013-01-01 05:00:00 2013-01-13 05:00:00
 5 Bikes    2013-01-01 05:00:00 2013-01-13 05:00:00
 6 Accesso~ 2013-01-01 05:00:00 2013-01-13 05:00:00
 7 Accesso~ 2013-01-01 05:00:00 2013-01-13 05:00:00
 8 Bikes    2013-01-01 05:00:00 2013-01-13 05:00:00
 9 Accesso~ 2013-01-01 05:00:00 2013-01-13 05:00:00
10 Accesso~ 2013-01-01 05:00:00 2013-01-13 05:00:00
# ... with 54,761 more rows, and 6 more variables:
#   ShipDate <dttm>, OrderQuantity <int>, UnitPrice <dbl>,
#   SalesAmount <dbl>, TaxAmt <dbl>, `Order Month` <chr>
```

Note how nicely the information printed. The tibble package only printed information that fitted nicely on the screen, and it gave information about the columns (variables) that were not displayed.

Step 9: Combine the code into one script and paste into the R editor for Power BI

Now it's time to take all the snippets from the prior steps and create a complete script that you can execute from Power BI. The script is shown in Listing 3-1, and you are now ready to paste it into the Power BI R editor.

Listing 3-1. The complete R script needed to combine the files

```r
library(tidyverse)
library(lubridate)
library(stringr)

setwd("./Data/SalesData")
monthly_reports <- list.files(".")
date_format <- stringr::str_replace(monthly_reports, ".csv","")
date_format <- lubridate::ymd(date_format)
df <- data_frame(monthly_reports, date_format)

max_month <- max(df$date_format)
min_month <- max_month %m+% months(-12)

reports_to_read <-
    df$monthly_reports[df$date_format >= min_month]

df_output <- purrr::map_df(reports_to_read, read_csv)
```

You copy the code in the R script and open the pbix file for this solution. Next, click *GetData* ➤ *More* ➤ *Other* to open the menu of connector options. Then click *R script* on the right-hand side of that menu as pictured in Figure 3-1.

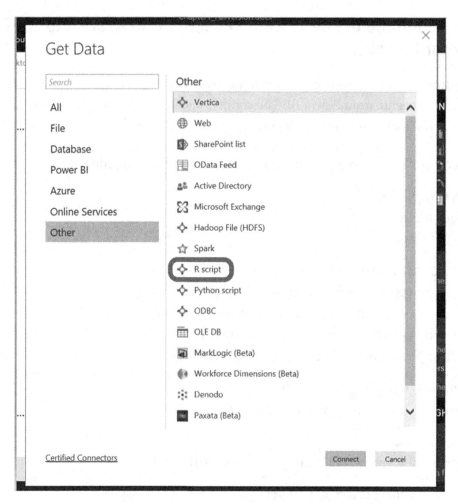

Figure 3-1. *Getting to the R script editor*

After you click *R script,* the editor for R in Power BI will appear which is pictured in Figure 3-2.

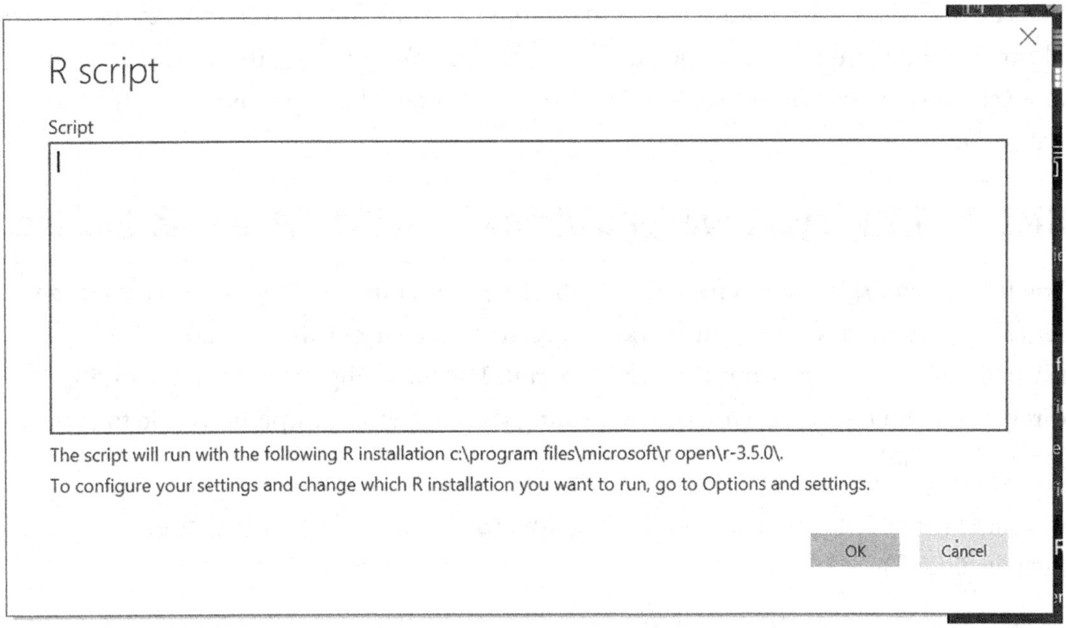

Figure 3-2. *The R script editor*

Now paste your R script from Listing 1-1, into the editor and click *OK*. Now you will be able to treat the resulting data set the same way you are able to treat any other data set that is brought into Power BI. You can make further transformations to it using Power Query, or you can load it straight into the Power BI data model.

Picking the rolling 24 months using Python

Like R, Python has been around since the early 1990s. It also has many tools and libraries that make the task of combining files over a rolling period relatively easy. Here are the steps to pick the files that are in the rolling 24 months using Python.

Step 1: Create a Python script and load the necessary libraries

The first thing you need to do is load the libraries and functions that will be used in the script. You do that using the following code:

```
import os
import pandas as pd
from dateutil.relativedelta import relativedelta
```

This code loads two libraries (*os* and *pandas*) and one function (*relativedelta*). The *os* library is used to interact with the file system, the *pandas* library is used to facilitate data analysis, and the *relativedelta()* function from the *dateutil* library is used to help with date manipulation.

Step 2: Change your working directory to the Python_Code folder

You use the *chdir()* function from the *os* library in the same fashion that you used the *setwd()* function in R. To set the working directory, you supply the function with a file path using the same rules that we used in R. The following code sets the working directory to the *Python_Code* folder assuming you saved the companion code to your *Documents* folder:

```
os.chdir("C:/Users/<"username">/Documents/AdvancedAnalyticsPowerBI/
Python_Code")
```

Now your script can use relative file paths to access the contents in the *SalesData* folder.

Step 3: Read the file names into a Python list

Now you read the names of the files located in the *SalesData* folder into a list. If you want to select the files that should be included in the rolling 24 months, you need to be able to analyze the file names to determine which files should be included. To do that, you will use the *os.listdir()* function from the *os* library to get all of the names of the files in the *SalesData* folder into a Python list. Here is the code you need to use to perform that task:

```
monthly_reports = os.listdir("./Data/SalesData/")
```

The following code tells Python to read all the file names in the *SalesData* folder into a Python list and assign the list to a variable named *monthly_reports*. Here is a printout of the contents contained in the *monthly_reports* variable:

```
>>> monthly_reports
['2010-12-01.csv', '2011-01-01.csv', '2011-02-01.csv',
 '2011-03-01.csv', '2011-04-01.csv', '2011-05-01.csv',
 '2011-06-01.csv', '2011-07-01.csv', '2011-08-01.csv',
 '2011-09-01.csv', '2011-10-01.csv', '2011-11-01.csv',
 '2011-12-01.csv', '2012-01-01.csv', '2012-02-01.csv',
```

```
'2012-03-01.csv', '2012-04-01.csv', '2012-05-01.csv',
'2012-06-01.csv', '2012-07-01.csv', '2012-08-01.csv',
'2012-09-01.csv', '2012-10-01.csv', '2012-11-01.csv',
'2012-12-01.csv', '2013-01-01.csv', '2013-02-01.csv',
'2013-03-01.csv', '2013-04-01.csv', '2013-05-01.csv',
'2013-06-01.csv', '2013-07-01.csv', '2013-08-01.csv',
'2013-09-01.csv', '2013-10-01.csv', '2013-11-01.csv',
'2013-12-01.csv', '2014-01-01.csv']
```

Step 4: Create a pandas data frame that will hold the information of the files to combine

The *pandas data frame* is a data structure that is similar to an R data frame. You will leverage a pandas *data frame* for this task. One of the easiest ways to manually create a *pandas data frame* is by converting a python *dictionary* to a *data frame*. You can think of a *dictionary* as python data structure that contains one or more *key/value* pairs.

You will create a dictionary with one *key/value* pair named *d,* and it will be based on the *monthly_reports* variable. The code you will use to do so is as follows:

```
d = {'monthly_reports': monthly_reports}
```

Next, you convert the *dictionary* to a pandas *data frame* using the following code:

```
df = pd.DataFrame(d)
```

The preceding code takes the dictionary *d* and uses it to create a one-column data frame. The *key* in dictionary *d* is used as the column name, and the *value* in dictionary *d* contains the items that will populate the rows.

Step 5: Create a new column that strips the date information from the monthly_reports column

Now you need a new column with just the date information. You create that new column using the following code:

```
df["date_format"] =pd.to_datetime(
    df.monthly_reports.str.replace(".csv","")
    )
```

This code first strips the string *.csv* from the file names in the *monthly_reports* column. Doing this leaves you with what is a valid date that is in a string format. To get the value in a date format, you use the *to_datetime()* function from the *pandas* library to parse the string to a date format.

Step 6: Get the upper and lower bound of the date range

Now you are in a position to get the upper and lower bound of the required date range. You are aiming for a rolling 24-month range. You can generate the required month range using the following code:

```
min_month = df.date_format.max() - relativedelta(months=23)
```

The logic used to calculate the *min_month* is to first find the max month of the files in the folder by getting the max of the *date_format* column. Notice you did so using *dot* notation to access the *max()* method of the *date_format* column.

Each column in a *pandas data frame* is of the *series* object type. The *series* object type includes built-in methods that you can use depending on the data type of the *series*. One of the methods that is available for *series* objects of the *datetime* data type is the *max()* method.

Now that you have the max date of the report files, you can calculate the date that is 24 months earlier than the max date. You do that by using the *relativedelta()* function from the *dateutil* library and passing *months=23* as a parameter. The result of using the *relativedelta()* function is that 23 months will be subtracted from the max date in the *date_format* column.

Step 7: Subset the data frame

Now that you know the date range, you can subset the *df* data frame to include only the rows that represent data in that range. You do so by using the following code:

```
reports_to_read = df[df["date_format"]>=min_month]
```

The code in the brackets creates a series of *Boolean* values based on whether or not the given date in the *date_format* column is greater than the *min_month* variable. Python uses the *True* values to subset the *df* data frame and assigns the results to the *reports_to_read variable*. Here's a printout of what the contents of the *reports_to_read* variable look like:

```
>>> reports_to_read
    monthly_reports date_format
14  2012-02-01.csv  2012-02-01
15  2012-03-01.csv  2012-03-01
16  2012-04-01.csv  2012-04-01
17  2012-05-01.csv  2012-05-01
18  2012-06-01.csv  2012-06-01
19  2012-07-01.csv  2012-07-01
20  2012-08-01.csv  2012-08-01
21  2012-09-01.csv  2012-09-01
22  2012-10-01.csv  2012-10-01
23  2012-11-01.csv  2012-11-01
24  2012-12-01.csv  2012-12-01
25  2013-01-01.csv  2013-01-01
26  2013-02-01.csv  2013-02-01
27  2013-03-01.csv  2013-03-01
28  2013-04-01.csv  2013-04-01
29  2013-05-01.csv  2013-05-01
30  2013-06-01.csv  2013-06-01
31  2013-07-01.csv  2013-07-01
32  2013-08-01.csv  2013-08-01
33  2013-09-01.csv  2013-09-01
34  2013-10-01.csv  2013-10-01
35  2013-11-01.csv  2013-11-01
36  2013-12-01.csv  2013-12-01
37  2014-01-01.csv  2014-01-01
```

Step 8: Combine files into one data frame

You need to create a data frame that is based on the union of all of the files that are in the rolling 24-month time period. You can accomplish this using the following code:

```
df_output = pd.concat(
    map(
        pd.read_csv,
        "./Data/SalesData/"+reports_to_read["monthly_reports"]))
```

The code uses the *map()* function in a similar fashion that you used the *map_df()* function from the *purrr* package in R. The map function in the preceding code maps the *read_csv()* function from pandas to each element in the *monthly_reports* series. That action results in a list of data frames. The *concat()* function from *pandas* is used to combine the list of data frames into one data frame.

Step 9: Add the code to Power BI

The complete code for this task is shown in Listing 3-2.

Listing 3-2. The complete Python script needed to combine the files

```
import os
import pandas as pd
from dateutil.relativedelta import relativedelta

os.chdir("<path to where the Python_Code folder is located>")
monthly_reports = os.listdir("./Data/SalesData")

d = {'monthly_reports': monthly_reports}
df = pd.DataFrame(d)
df["date_format"] = pd.to_datetime(
    df.monthly_reports.str.replace(".csv",""))

min_month = df.date_format.max() - relativedelta(months=23)
```

```
reports_to_read = df[df["date_format"]>=min_month]

df_output = pd.concat(
    map(
        pd.read_csv,
        "./Data/SalesData/"+reports_to_read["monthly_reports"]))
```

You copy the code in the Python script, then open up our pbix file for this example. Next, you click *GetData* ➤ *More* ➤ *Other* and select *Python script* on the right-hand side. Figure 3-3 shows the screen where you will find Python script editor.

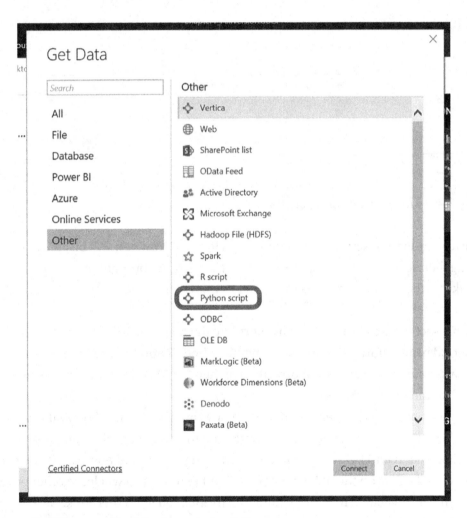

Figure 3-3. *Where to find Python script editor*

After you perform that action, the Python editor for Power BI will appear as shown in Figure 3-4.

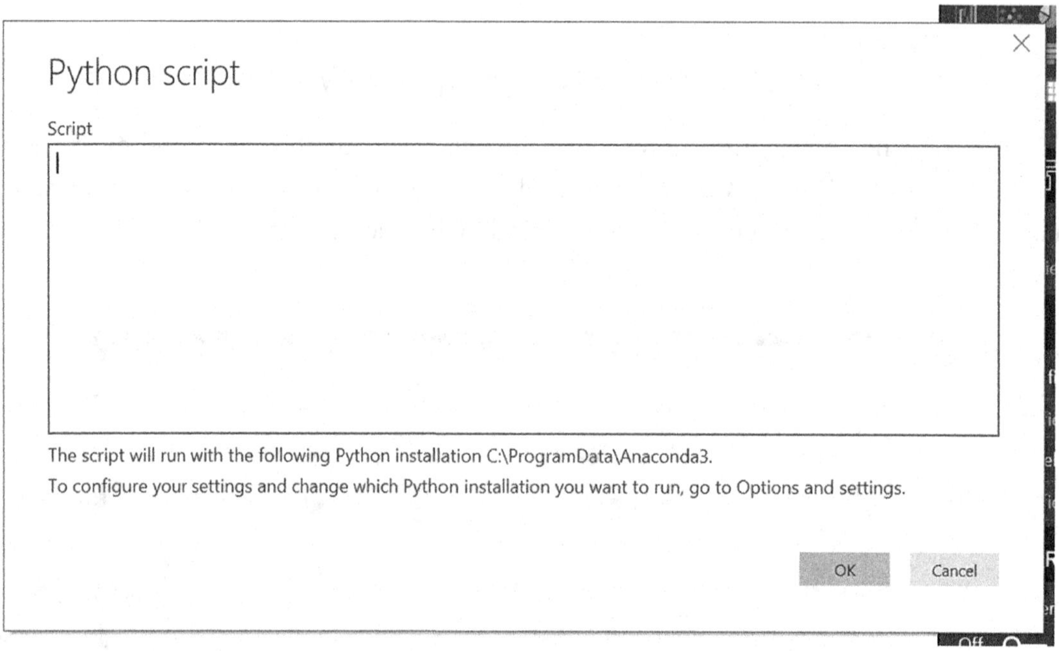

Figure 3-4. *The Python script editor*

Paste your Python script in this editor and click *OK*. Now you will be able to treat the resulting data set the same way you treat any other data set brought into Power BI. You can make further transformations to it using Power Query, or you can load it straight into the data model.

In this section, we learned how to determine the files that are in the rolling 24-month time period using information in the file name in both R and Python. The code needed to do it was straightforward and more intuitive than the M code would have been to accomplish the same task.

You may never need to perform this exact task, but you can modify what you learned in this section to fit your needs. I want you to open your mind and look at the possibilities that R and Python give in Power BI. You just learned how easy it is for you to interact with the files on the file system with R and Python. You will learn other powerful features that are available in R and Python the further you go in this book. Next, you will learn how to leverage regular expressions in R and Python to filter rows from a *csv* file.

Filtering rows based on a regular expression

Power BI enables you to remove records that you don't want to bring into your data model fairly easy for most situations, but there are some situations where the logic that is needed to filter the data is not possible using native Power BI tools. One example is filtering records in your data set based on a string pattern that can only be matched with a *regular expression*. You cannot leverage *regular expressions* in Power Query, but you can in both R and Python.

In the next scenario, let's pretend that you are the chief analyst for a political campaign. You have a file that contains a list of potential voters that you need to analyze for an email campaign. You only want to include voters in your Power BI-based analysis that have a well-formed email address. You will use a *regular expression* to detect whether the email addresses in the data set are well-formed. So in your analysis, a record with an email address of johndoe@email will be discarded, but a record with an email address of janedoe@email.com will be kept. The result of your scrubbing will be a data set with only well-formed email addresses that will be brought into the Power BI data model.

Leveraging regular expressions via R

As of the writing of this book, this task would be impossible to do using Power Query because *regular expressions* are not available in Power Query. That is not the case in R because *regular expressions* are first-class citizens in that programming language. The code needed to accomplish this task in R is concise and intuitive. Here are the steps.

Step 1: Load the required packages

The first thing you need to do is load the required packages so that you can perform the task in R. The two packages you need are *tidyverse* and *stringr*. You import the required packages using the following code:

```
library(tidyverse)
library(stringr)
```

Note that *stringr* is part of *tidyverse,* but it is not part of the core packages that are loaded when you load the *tidyverse* metapackage so you need to explicitly load it.

Step 2: Load the file that contains the potential voters into R

You accomplish the task using the following code:

```
setwd("path to where the R_Code folder is located")
goodemails_raw <- read_csv("./Data/EmailAddresses.csv")
```

The code is straightforward. You first set the working directory to the *R_Code* folder using the *setwd()* function. Next, you use *relative file path referencing* to traverse to the *EmailAddresses.csv* file.

Step 3: Define the regular expression

You use the following code to define the required regular expression:

```
email_pattern <- "^([a-zA-Z][\\w\\_]{4,20})\\@([a-zA-Z0-9.-]+)\\.
([a-zA-Z]{2,3})$"
```

The preceding regular expression is what we will use to match the pattern of a well-formed email address. Please note that the preceding regular expression represents one line of code, but it is occupying two lines due to width limitations of the page. Don't be alarmed with the complexity of regular expressions. You can often find the regular expression you need for a given task via a *Bing* search. Remember, *"Good programmers write good code; great programmers steal great code"* (unknown author). However, I will not call using a regular expression that you find via an Internet search "stealing" because the altruistic nature of many developers makes them want to share their knowledge with the community.

Step 4: Remove the bad email addresses from the data set

You will do so using the following code:

```
goodemails <-
    goodemails_raw %>%
    filter(str_detect(Email, email_pattern) == TRUE)
```

The code uses the *filter()* function from the *dplyr* package to filter the *goodemails_raw* data frame. The filtering is done using the *str_detect()* function from the *stringr* package. The *str_detect()* function enables you to test whether a string matches a regular expression pattern. It returns *TRUE* if it matches the regular expression pattern passed to it; otherwise, *FALSE* is returned. The *filter()* function will only retain rows where the *str_detect()* function evaluates to TRUE.

Step 5: Combine the preceding code into one script and paste in the R editor for Power BI

The complete script is in Listing 3-3.

Listing 3-3. The complete R script needed to scrub bad emails

```
library(tidyverse)
library(stringr)

setwd("<full path to the folder that contains the companion code for the
chapter>")

goodemails_raw <- read_csv("./Data/EmailAddresses.csv")

email_pattern <- "^([a-zA-Z][\\w\\_]{4,20})\\@([a-zA-Z0-9.-]+)\\.([a-zA-Z]
{2,3})$"

goodemails <-
        goodemails_raw %>%
        filter(str_detect(Email, email_pattern) == TRUE)
```

Copy the code in the R script and open your pbix file for this solution. Next, click *GetData* ➤ *More* ➤ *Other,* then select *R script* on the right-hand side. Click *OK* after you paste the code into the *R editor* in Power BI. The resulting data set will be added to the Power BI data model.

Leveraging regular expressions via Python

In this exercise, you will leverage *regular expressions* in Power BI using Python. Just like with R, the Python code needed to accomplish this task is concise and intuitive. Here are the steps.

173

Step 1: Load the required libraries

The first thing you need to do is load the required libraries for the task. The two libraries we need are *os* and *pandas*. You load the libraries using the following code:

```
import os
import pandas as pd
```

Step 2: Load the file that contains the potential voters into Python and assign the contents to a pandas data frame

You accomplish the task using the following code:

```
os.chdir("<path to the Python_Code folder in Chapter 3>")
goodemails = pd.read_csv("<full path to EmailAddresses.csv>")
```

First, you change the working directory to the *Python_Code* folder, then you read in the contents of the *EmailAddresses.csv* file into a pandas data frame, and assign the results to the *goodemails_raw* variable.

Step 3: Define the regular expression that matches the pattern of a well-formed email address

You use the following code to do so:

```
email_pattern <- "^([a-zA-Z][\\w\\_]{4,20})\\@([a-zA-Z0-9.-]+)\\.
([a-zA-Z]{2,3})$"
```

The preceding regular expression is what you use to test the email addresses. The code is wrapped on two lines, but it actually represents just one line of code. It is on two lines due to width limitations of the book. Note that the same regular expression that you used in R also works in Python. Don't be alarmed with the complexity of regular expressions. You can often find the regular expression you need for a given task via a *Bing* search.

Step 4: Remove the bad email addresses from the data set

You will remove rows that contain bad email addresses using the following code:

```
goodemails = goodemails_raw[
    goodemails_raw["Email"].str.match(email_pattern)]
```

The code works by first creating a Boolean pandas series using the goodemails["Email"].str.match(email_pattern) code that is located between the opening and closing brackets. That Boolean series is used to filter the data frame by only keeping the rows where the code evaluates to *True*.

Step 5: Combine the preceding code into one script and paste in the Python editor for Power BI

The complete script is shown in Listing 3-4.

Listing 3-4. The complete Python script

```
import os
import pandas as pd

os.chdir("C:/Users/ryanwade44/Downloads/Advanced Analytics in Power BI with
R and Python/Chapterx03/R_Code")
goodemails_raw = pd.read_csv("./Data/EmailAddresses.csv")

email_pattern = "^([a-zA-Z][\\w\\_]{4,20})\\@([a-zA-Z0-9.-]+)\\.
([a-zA-Z]{2,3})$"

goodemails = goodemails_raw[
    goodemails_raw["Email"].str.match(email_pattern)]
```

Now you copy the code in Listing 3-4. and open up your pbix file for this solution. Next, you click *GetData* ➤ *More* ➤ *Other,* then select *Python script* on the right-hand side. After that, you paste the code into the *Python editor,* then click *OK* which adds the data set to the Power BI data model.

As illustrated in this example, regular expressions are extremely useful when it comes to pattern matching strings. They are a very valuable tool for all data analysts to have in their toolkit. They are not available for use in Power Query, but fortunately, they are in Python and R. This is just one example how regular expressions can help you accomplish pattern matching tasks that are not possible using the Power Query string functions. Keep reading and you will be introduced to more examples of how you can use *regular expressions* later in the book. Next, we will learn how to read MS Excel files into the Power BI data model.

CHAPTER 4

Reading Excel Files

Power BI has built-in tools that enable you to read Microsoft Excel files into the Power BI data model. The built-in functionality is intuitive and easy to use for simple workflows, but things can become unnecessarily difficult when the workflows get complicated. Many workflows that are difficult to do using native Power BI tools are relatively easy to do using R or Python. We will illustrate with an example.

Here is the scenario:

- It is February 2, 2014, and you are the CFO of Contoso International.

- You receive sales data by year via MS Excel workbooks from your IT department, and you save the data in a folder named *ExcelFiles*.

- Each workbook is broken out by year, and each month has its own sheet in the workbook.

- The *ExcelFiles* folder contains five workbooks in total. The workbook for calendar year 2010 contains sales data just for December of that year, the workbooks for calendar years 2011–2013 have a full year worth of data, and the workbook for calendar year 2014 has data just for January of that year.

- At the beginning of each month, a new sheet is added to the workbook with the sales data from the previous month.

- At the beginning of February of each year, you get a new workbook for the current year that is updated in the same manner as the previous workbooks.

© Ryan Wade 2020

R. Wade, *Advanced Analytics in Power BI with R and Python*, https://doi.org/10.1007/978-1-4842-5829-3_4

You need to combine these files so that you can perform your required financial analysis. You tried to automate this using Power Query but was not successful because it required advanced Power Query techniques that was too difficult to understand, so you decided to try to develop a solution using R and Python. Your friend, Tyrone, is proficient in both R and Python, so he was able to help you. He gave you the steps of how to accomplish the task in both R and Python. We will go over the R example first.

Reading Excel files using R

In this example, you will create a workflow that

1. Loops through each workbook in the ExcelFiles folder

2. Combines the data from every sheet in the workbook into one data frame

3. Appends the data frame created in Step 2 to the master data frame that will warehouse the data from all of the workbooks when the workflow is completed

The image in Figure 4-1 provides a visual representation of the steps.

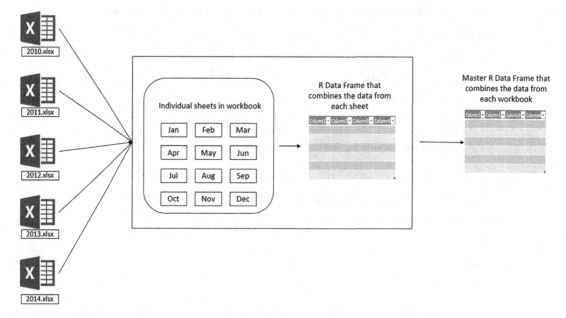

Figure 4-1. *Workflow to combined total sales in an R data frame*

Now let's go over the steps needed to convert the preceding workflow into R code.

Step 1: Import the tidyverse and readxl package

The packages that are used for this workflow are *tidyverse* and *readxl*. You are already familiar with *tidyverse*. The *readxl* package is similar to the *readr* package but is specifically designed to work with Microsoft Excel workbooks. Here is the required code to load the packages:

```
library(tidyverse)
library(readxl)
```

Step 2: Create the shell of the combine_sheets function

In this step, you are creating the beginning of a function that will read in the data that are contained in the worksheets of the given workbook and combine them into one data frame. The file path of the Excel workbook that you want to operate on is passed to the function. Here is what the shell of that function looks like:

```
combine_sheets <- function(excel_file_path) {
    <R Code>
    return(<R object>)
}
```

Step 3: Get the name of the sheets you need to combine in your function from the specified Excel workbook

Now you need to start defining the body of your function. Your function needs to identify the sheets in your workbook that it needs to combine. It will do so using the following code:

```
df <- excel_file_path %>%
        excel_sheets()
```

The preceding code starts a workflow using the file path of the Excel workbook that is contained in the *excel_file_path* variable. The workbook at that path contains the sheets you want to combine. The *excel_file_path* variable is passed to the *excel_sheets()* function from *readxl* using the pipe (%>%) operator. The *excel_sheets()* function returns a character vector that contains the names of the sheets in the workbook located in the path specified in the *excel_file_path* variable. This function relies heavily on the %>% operator. Refer to the book's introduction if you need an explanation of what it is and how to use it.

Step 4: Convert the character vector built in Step 3 to a named character vector

You convert the character vector created in Step 3 into a *named character vector*. The name property of the *named character vector* will be used to identify the worksheet where the rows of the resulting data frame produced in this example came from. The *named character vector* is created using the following code:

```
df <- excel_file_path %>%
    excel_sheets() %>%
    set_names() %>%
```

The preceding code takes the character vector that was created by the *excel_sheets()* function and passes it to the *set_names()* function from the *purrr* package. This function sets the names of each element in the character vector to itself. You can get a better understanding of this function by inspecting the *df* variable before and after the *set_names()* function is applied. Here are the contents of the *df* variable before *set_names()* is applied:

```
> df
 [1] "Jan" "Feb" "Mar" "Apr" "May" "Jun" "Jul" "Aug" "Sep" "Oct" "Nov" "Dec"
```

Here are the contents of the *df* variable after the *set_names()* function is applied:

```
> df
  Jan    Feb    Mar    Apr    May    Jun    Jul    Aug    Sep    Oct    Nov    Dec
"Jan"  "Feb"  "Mar"  "Apr"  "May"  "Jun"  "Jul"  "Aug"  "Sep"  "Oct"  "Nov"  "Dec"
```

Notice that each element in the character vector has been given the name that is equal to the value of the element. You will take advantage of this property in the next step.

You are not limited to naming the character vector to itself. You can pass a character vector of the same length with the names you want to use or a function for situations where a name already exists, and you want to modify it based on some rule.

Step 5: Use the mapr_df function to combine the sheets into a single data frame

In this step, you add one line of code to the workflow, but there is a lot going on in this line of code that needs to be unpacked. Here is what the workflow looks like with the extra code added:

```
df <- excel_file_path %>%
      excel_sheets() %>%
      set_names() %>%
      map_dfr(.f= ~read_excel(path= excel_file_path, sheet= ..1)
             ,.id = "sheet"
      )
```

Let's explain what's going on in the last line:

- The named character vector created in the previous step is passed to the *map_dfr()* function as the first argument.

- The *.f* argument is used to tell *map_dfr()* the function you want to apply to the named character vector that you passed to it. The function in this situation is *read_excel()*. You describe how you want to use the *read_excel()* function via a *formula*. The *formula* starts with a ~, then the *read_excel()* function along with its arguments. The arguments that it uses are the *path* and *sheet* arguments. The file path of the workbook that contains the sheets you want to combine is assigned to the *path* argument, and the value from the

named character vector is assigned to the *sheet* argument. The value
assigned to this argument was *..1*. This method is used to get the
argument we want to use based on its position. The named character
vector was passed to the *map_dfr()* function as the first argument via
set_names() so that is why we used *..1*. If it was in the *nth* position,
then we would set the value to *..n*.

- The *map_dfr()* function combines the data frames that were created
  for each sheet in the workbook and collapse it into one data frame.
  You can add a field to the resulting data frame that will enable you
  to determine the sheet that the record came from. You accomplish
  this using the *.id* argument. If you use this option, a column is added
  that uses the names from the *named character vector* to populate
  it. If you are not using a named character vector, the position of the
  sheet name in the character vector is used. The value you give this
  argument will be used as the column name.

Step 6: Return the data frame

You want your function to return the data frame that was created. To do so, you use the
return statement with the *df* data frame you created. After you do that, the complete
definition of the function looks like this:

```
combine_sheets <- function(excel_file_path) {
    df <- excel_file_path %>%
        excel_sheets() %>%
        set_names() %>%
        map_df(.f = ~read_excel(path = excel_file_path, sheet = ..1)
                ,.id = "sheet"
        )

    return(df)
}
```

Step 7: Set the working directory to the location where the Excel files are located

Now that you have the function defined, the hard part is over. You need to write the code that utilizes the custom *combine_sheets()* function to combine the data from the workbooks in the *ExcelFiles* folder. The first thing you need to do is set the working directory so Power BI knows what folder to use. You can do so using the following code:

```
setwd("<full path to the ExcelFiles folder>")
```

Step 8: Assign the file names to the excel_file_paths variable

You use the *list.files()* function to get the name of the files in the current directory and assign it to a variable named *excel_file_paths* using the following code:

```
excel_file_paths <- list.files(".")
```

The "." passed to the *list.files()* function tells it that you want to return all the files located in the current working directory. You can specify a nested folder in the following way "./<nested folder name>". The list.files() function also accepts a character vector containing one or more full path names of the paths you want to inspect.

Step 9: Use the map_dfr function to apply the combine_sheets function to each file in the working directory

Now you are at the step that combines the data from all the workbooks into one data frame. The code needed to do it is as follows:

```
combine_workbooks <- map_dfr(excel_file_paths, combine_sheets)
```

The *map_dfr()* function applies the *combine_sheets()* custom function to the Excel file names contained in the *excel_file_paths* character vector. The result of doing so is a list of data frames. The *map_dfr()* function collapses the list of data frames into one data frame and names the data frame *combine_workbooks*.

Step 10: Copy the R script and paste it into the R editor via GetData in Power BI

The resulting script of the previous steps is as follows:

```
library(tidyverse)
library(readxl)

combine_sheets <- function(excel_file_path) {
    df <- excel_file_path %>%
        excel_sheets() %>%
        set_names() %>%
        map_dfr(.f = ~read_excel(path = excel_file_path, sheet = ..1)
                ,.id = "sheet"
        )

    return(df)
}

setwd("<full path to the ExcelFiles folder>")
excel_file_paths <- list.files(".")
combine_workbooks <- map_dfr(excel_file_paths, combine_sheets)
```

After you paste the code in the R editor in Power BI, the resulting data set will be exposed to you like any other data set brought into Power BI.

Reading Excel files using Python

Now you will perform the same task in Python. The workflow will be similar, but the execution will be slightly different. The image in Figure 4-2 visually depicts the workflow that will be used in Python.

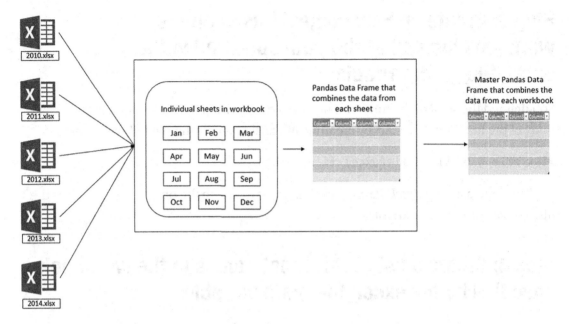

Figure 4-2. *Workflow to combined total sales in a Pandas data frame*

Step 1: Import the os and pandas library

The libraries that are used for this workflow are *os* and *pandas*. The *os* library is used to interact with the file system, and *pandas* is used to read data from Excel into Python.

Step 2: Create the shell of the combine_sheets function

You need to create a custom function that reads in the contents from the sheets in the Excel workbook and combine them into one data frame. You need to pass to the function the file path of the Excel workbook that contains the sheets you want to combine. Here is the shell of the custom function you will use to perform this task:

```
def combine_sheets(excel_file_path):

    <Python Code>

    return <Python object>
```

Step 3: Create an Excel object based on the workbook located at the path specified in the excel_file_path variable

You defined the shell of the function; now let's start writing the code that will go in the body of the function. You start by creating an Excel object using the following code:

```
xlsx_file = pd.ExcelFile(excel_file_path)
```

The code uses the *ExcelFile* class from *pandas* to create the object. All it needs is the file path to the Excel workbook.

Step 4: Create a list of the sheet names in the workbook specified by the excel_file_path variable

The function needs to know what sheets to combine. You can get that information using the *sheet_names* property of the *xlsx_file* object you created in *Step 3*. You do so using the following code:

```
ws = xlsx_file .sheet_names
```

The code gets the sheet names from the *sheet_names* property of *xlsx_file* and stores them in the *ws* variable.

Step 5: Use the read_excel method of pandas to read the data from each sheet into one data frame

You accomplish *Step 5* using the following line of code:

```
df = pd.concat(pd.read_excel(xlsx_file, sheet_name=ws))
```

Two major tasks are happening in the code, so it will be easier to understand if we unpacked it into multiple steps. The first thing that is done is *pandas* is using *pd.read_excel(xlsx_file, sheet_name=ws)* to read the data from the sheets in the workbook, and it puts them in individual data frames. The first argument is unnamed, but it is the *io* argument. This is where you identify the MS Excel file that you want to read. The second argument is the *sheet_name* argument, and that is where you tell the *read_excel()* method the worksheets you want to read from the specified workbook.

186

The *ws* variable is used to give *read_excel()* that information. The second thing *pandas* does is it combines the individual data frames that were created using the *read_excel()* method into one data frame via the *concat()* method from *pandas*. The resulting data frame is assigned to the variable named *df*.

Step 6: Return the data frame held in df as the output from the combine_sheets function

This is done using the return statement as shown here:

```
return df
```

Step 7: Set the working directory to the location that contains the Excel data you want to combine

You now have the *combine_sheets()* custom function completely defined. Now you need to write code that uses the function to combine the worksheets from the workbooks in the *ExcelFiles* folder into one data frame. You first need to set the working directory to the location that contains the Excel workbooks. You do so using the following code:

```
os.chdir("<full path to the ExcelFiles folder>")
```

Step 8: Get the list of the file names in the current working directory and assign it to the excel_file_paths variable

Your script needs to know the names of the MS Excel workbooks to combine. You can get the names of the workbooks using the *listdir()* function as illustrated in the following code:

```
excel_file_paths = os.listdir(".")
```

The code makes the assumption that only the Excel files you want to combine are in the current folder. That needs to be verified before the script is run.

Step 9: Create an empty data frame and name it combined_workbooks

In this step, you create an empty data frame named *combined_workbooks* that will be the master data frame that will store the data from all of the workbooks. The code needed to create the empty data frame is as follows:

```
combined_workbooks = pd.DataFrame()
```

The data frame created for each workbook using the combined_sheets function will be appended to this data frame.

Step 10: Create the shell of the for loop

The for loop in python enables you to iterate over a sequence. In this case, the sequence that will be iterated over is the *excel_file_paths* list. The script uses the following code to iterate over the *excel_file_paths* list:

```
for excel_file_path in excel_file_paths:
```

You will use *for* loops a lot in Python, so let's take some time to explain the structure. The *for* loop can be described using the following template:

```
for <variable to hold current value> in <sequence object>:
```

The variable that holds the current value of the iteration can be named anything. Typically, the name is the singular version of the name given to the sequence object. The sequence object is the object that you are iterating. In this script, it is the *excel_file_paths* list. Let's walk through the steps to explain how it operates.

If you print out the contents of *excel_file_paths,* you would see the following data:

```
['2010.xlsx', '2011.xlsx', '2012.xlsx', '2013.xlsx', '2014.xlsx']
```

You see that *excel_file_paths* is a list that contains five elements. The *for* loop in the script iterates over this list, and it assigns the value of *'2010.xlsx'* to the *excel_file_path* variable after the first iteration, assigns the value of *'2011.xlsx'* to the *excel_file_path* variable after the second iteration, and continues doing that for each element in the list until it iterates all of the elements.

Step 11: Combine all the data in each sheet into one data frame using the combine_sheets function

The following code takes the file path in the *excel_file_path* variable and passes it to *combine_function()*:

```
combined_workbook = combine_sheets(excel_file_path)
```

The *combine_sheets()* custom function combines the sheets in the workbook located at the file path that was passed to it and puts them into one data frame. That data frame is stored in the *combined_workbook* variable.

Step 12: Append the combined_workbook data frame to the combined_workbooks data frame

The code needed to perform this task is as follows:

```
combined_workbooks = combined_workbooks.append(combined_workbook,
                    ignore_index=True)
```

The code uses the append method of the *combined_workbooks* data frame to append the *combine_workbook* data frame to it. The *ignore_index* argument is set to *True* because you want to ignore the indexes in *combined_workbook* data frame and use the ones that are in *combined_workbooks*.

Step 13: Copy the Python script and paste it into the Python editor via GetData in Power BI

Here is the complete script:

```
import os
import pandas as pd

def combine_sheets(excel_file_path):

    xlsx_file = pd.ExcelFile(excel_file_path)
    ws = xlsx_file.sheet_names
    df = pd.concat(pd.read_excel(xlsx_file, sheet_name=ws))

    return df
```

```
os.chdir("<full path to SalesData_WorkbookFormat")
excel_file_paths = os.listdir(".")
combined_workbooks = pd.DataFrame()

for excel_file_path in excel_file_paths:
    combined_workbook = combine_sheets(excel_file_path)
    combined_workbooks = combined_workbooks.append(
                        combined_workbook, ignore_index=True)
```

If you copy the script and paste it to the Python editor in Power BI via *GetData*, you will get a data set that you can add to the Power BI data model.

In this chapter, you were able to implement a workflow that would have been very difficult to do using Power Query and M. The R and Python code needed to do the workflow is more succinct and easier to understand than the M code that is necessary to perform the same task. You also learned how to use R and Python to abstract logic that is repeated multiple times in a function to improve code efficiency and readability. In the next chapter, you will learn how to interact with *SQL Server* in Power BI via R and Python.

CHAPTER 5

Reading SQL Server Data

Power BI's built-in ETL tool, Power Query, is capable of reading data that is stored in a variety of formats into memory, making transformations to the data while in memory, and loading the transformed data into the Power BI data model. It does a great job of doing these types of tasks as it can handle most ETL situations.

However, Power BI falls short if you want to log information about your Power BI data loads. Power Query is designed to read data but not to write data. Fortunately, that is not the case with R and Python. R and Python have packages and libraries, respectively, that make logging data relatively easy. Both R and Python can write to multiple sources ranging from flat files to databases. In this example, you will use R and Python to log information about your data loads to a SQL Server database.

Adding AdventureWorksDW_StarSchema database to your SQL Server instance

The example in this chapter requires the *AdventureWorksDW_StarSchema* database as the data source. This database is a subset of the *AdventureWorksDW* database developed by Microsoft. Perform the following steps to restore the database to your SQL Server instance:

1. Go to the repo of the book, then traverse to the *Data* folder for Chapter 5. You will find the zip version of the *AdventureWorksDW_StarSchema.bak* file in this location. Unzip the *AdventureWorksDW_StarSchema.zip* to expose *AdventureWorksDW_StarSchema.bak*.

2. Copy the *AdventureWorksDW_StarSchema.bak* file to the *Backup* folder for SQL Server. The location of this folder in the *Data Science Virtual Machine (DSVM)* is C:\Program Files\Microsoft SQL Server\MSSQL15.MSSQLSERVER\MSSQL\Backup.

© Ryan Wade 2020

R. Wade, *Advanced Analytics in Power BI with R and Python*, https://doi.org/10.1007/978-1-4842-5829-3_5

3. Go to *SQL Server Management Studio* (SSMS) and right-click the *Databases* folder, then select *Restore Databases....*

4. The *Restore Database* pop-up form will appear. Make sure you are on the *General* page tab. In the *Source* section, select the *Device* radio button.

5. Click the eclipse button, the button with the three dots, then click the *Add* button. Next, browse to the location where *AdventureWorksDW_StarSchema.bak* is located.

6. Select the *AdventureWorksDW_StarSchema.bak* file, then click the *OK* button to close out the *Locate Backup File* form.

7. Click *OK* to close out the *Select backup devices* form.

8. Click *OK* to close out the *Restore Database* form. Doing so will add the database.

Reading SQL Server data into Power BI using R

Here is the scenario. You are an analyst for *Clothing R Us*. The company has a small IT department, and the database administrator, Vihaan, doubles as the system engineer. There is no dedicated staff for business intelligence. You convinced the CFO, Adaugo, about the advantages of business intelligence and how it can help the business make informed decisions based on the company's experience. She agreed to let Vihaan build a small data mart around Internet sales. The data mart contains one fact table and four dimension tables. All the dimension tables are Type 1 dimensions, and each load of the warehouse is a *kill and fill*. Vihaan developed an ETL process, but the ETL did not include any logging to monitor the amount of data that is being loaded into the data mart. The success of this POC is critical to you, so you took it upon yourself to develop a logging system as a sanity check to make sure that the data that is being loaded into Power BI is reasonable.

In Type 1 Slowly Changing Dimension (SCD), the new information overwrites the original information so no history is kept. In Type 2 SCD, a new record is added to the table to represent the new information. Both the original and the new record will be present.

You know that Power Query is designed to read data into Power BI, but it is not designed to write data. Your colleague, Sarah, informed you that you can write data back to a database relatively easy using either R or Python. You shared with her what information you needed to log, and she gave you instructions for both R and Python. Here are the instructions for R.

Step 1: Create a DSN to the SQL Server database

You need to provide SQL Server with information to authenticate you so that you can connect to the database. There are multiple ways to do this, but the method you will use is the DSN (Data Source Name) method. DSNs are relatively easy to create and configure in Windows 10. Here are the steps to do so:

1. Go to the *Cortana Search Bar* located on the left side of the taskbar next to the *Windows* icon and type *administrative tools.*

2. You will see a list of administrative tools shortcuts. Search for *ODBC Data Sources (64-bit)* and double-click it.

3. A pop-up box for *ODBC Data Source Administrator (64-bit)* should appear with multiple tabs. Click the *System DSN* tab. You want to use this tab to ensure that all users on your computer will have access to this DSN. If you only want the current user to have access to the DSN, you can stay on the *User DSN* tab.

4. Click the *Add* button.

5. You will see a list of drivers that you can choose from with SQL Server being one of them. Select *SQL Server,* then click the *Finish* button.

6. Next the *Create a New Data Source to SQL Server* pop-up box will appear. This is where you configure the DSN. The *Name* is the name that will be used by R to refer to the DSN, the *Description* can be anything you want to use to describe the DSN, and the *Server* is the name of the SQL Server that warehouses the database you want to connect to.

Use *SQLServer2019* for the *Name* and SQLServer2019 for the *Description*. If you are not using SQL Server 2019, then use a name that identifies the version of SQL Server you are working with. The image below shows an example in which SQL Server 2017 was used. Next, you need to retrieve the name of the SQL Server instance that you want to use. To get the server name, open *SQL Server Management Studio* and connect to the SQL Server instance. The name that appears in the *Server Name* combo box is the name that you want to use in the *Server* combo box in the *Create a New Data Source to SQL Server* form. Copy the server name and paste it in the *Server* combo box. When you finish filling out the form, it should look like the image in Figure 5-1. Click *Next* if everything looks correct.

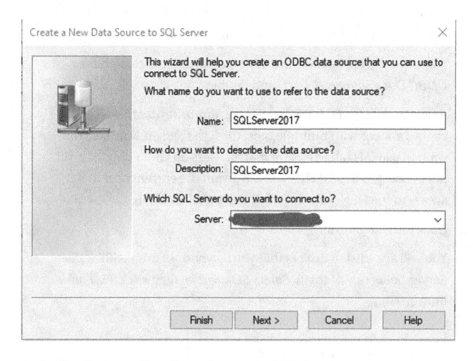

Figure 5-1. *The Create a New Data Source to SQL Server*

7. You will see the pop-up form shown in Figure 5-2. Make sure yours is configured the same way, then click *Next*.

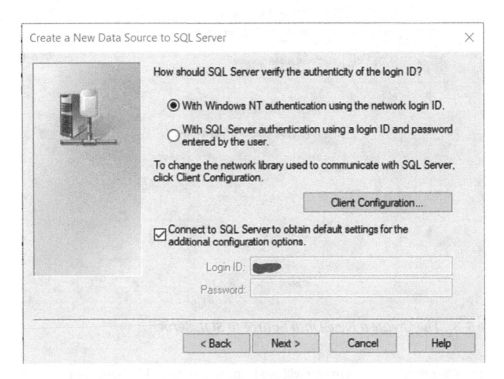

Figure 5-2. *The Create a New Data Source to SQL Server*

8. Next, you will see the pop-up form in Figure 5-3. Make sure you check the *Change the default database to:* check box and change the database to *AdventureWorksDW_StarSchema,* then click *Next*.

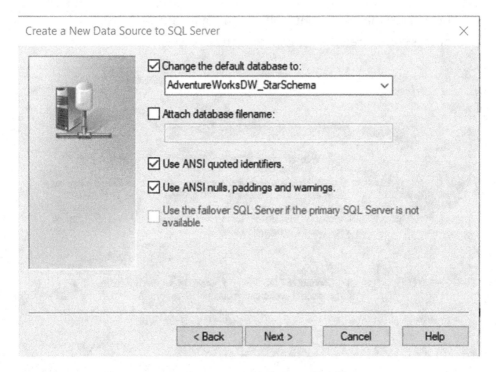

Figure 5-3. *The Create a New Data Source to SQL Server*

9.   The next pop-up form you will see is the one depicted in Figure 5-4.
      Make sure yours is configured the same way, then click *Finish*.

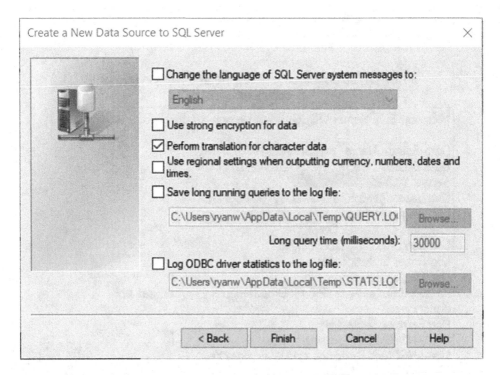

Figure 5-4. The Create a New Data Source to SQL Server

10. After you click *Finish,* the pop-up box like the one shown in Figure 5-5 will appear. You can test the connection by clicking the *Test Data Source* button to make sure you can successfully connect to the database using the *DSN* name you just configured.

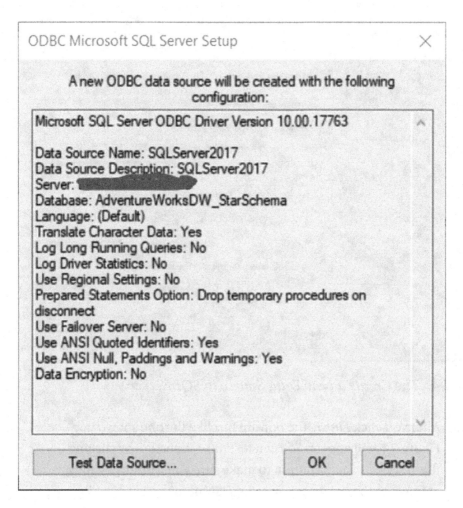

Figure 5-5. *Information about the DSN*

Step 2: Create a log table in SQL Server

A log table with the name dbo.LoadHistoryLog should already exist in the database. In case you need to recreate it, you can do so by executing the following code in SQL Server Management Studio (SSMS):

```
USE AdventureWorksDW_StarSchema

CREATE TABLE dbo.[LoadHistoryLog] (
    DATESTAMP DATETIME,
    TABLENAME VARCHAR(25),
    NUM_RECORDS INT
)
```

To execute the code in SSMS, you copy the preceding code, then click the *New Query* button located in the menu bar. An editor should appear. Paste the code in the editor, then click the *Execute* button in the menu bar to execute the code.

Step 3: Start developing the R script to load DimDate

Go back to R Studio to begin developing the R script. The packages that you will use in this script are *RODBC* and *lubridate*. The *RODBC* package will be used to interact with SQL Server, and the *lubridate* package will be used to make it easier to do some date-related tasks. Here's the code needed to load the packages:

```
library(RODBC)
library(lubridate)
```

Step 4: Create a variable to hold the name of the table you want to import

The name of the table that will be read into the Power BI data model will be used in several places in the script. A variable is used instead of hard-coding the value throughout the script to make the script easier to maintain and reuse:

```
table_name <- "DimDate"
```

Step 5: Create a variable to hold the sql statement that will be used to return the table

The following code is used to create the sql statement that will be used to read the data from the DimDate table into Power BI:

```
sql <- paste0(
    "SELECT * FROM dbo.",
    table_name)
```

The preceding code is using the *paste0()* function to concatenate "SELECT * FROM dbo." with the *table_name* variable. The *paste0()* function is similar to the *paste()* function. The difference between the two is that you can specify a delimiter in the *paste()* function but the *paste0()* function assumes that you want to concatenate strings without using a delimiter.

Step 6: Create a connection to SQL Server

You create a connection to the database using the following code:

```
conn <- odbcConnect(dsn = "SQLServer2017")
```

As you can see, using a DSN makes it very easy to connect to the database. You use the *odbcConnect* function from the *RODBC* package to make the connection, and the only argument you need is the *dsn* argument. You set that argument equal to the name of the *DSN* that you configured in Step 1.

Step 7: Retrieve the data from SQL Server and store it in a data frame

The *sqlQuery()* function from *RODBC* enables you to submit a query to the database using the *conn* connection and stores the results in a data frame. The code to perform this operation is this:

```
df_read <- sqlQuery(channel = conn, query = sql)
```

The results of the query are stored in the *df_read* data frame. This is the data that will be fed to Power BI.

Step 8: Get the current time

Now you need to start calculating the information you want to log about the load. The information you want to log is the time the load occurred, the name of the table you loaded, and the number of records that were loaded.

To get the time of the load, you use the *now()* function from the *lubridate* package and assign it to the *datestamp* variable using the following code:

```
datestamp <- now()
```

Step 9: Get the number of records that were read

You want to log the number of records that were loaded from the *DimDate* table. You get that information using the *nrow()* function from base R. You use the following code to store the number of records in DimDate to the *num_records* variable:

```
num_records <- nrow(df_read)
```

Step 10: Create a one record R data frame that contains the information you want to log

You calculated the information that you want to log; now you need to shape it so that you can insert the information into the log table. The shape the data needs to be in is a data frame. You start off by creating a *named list* using the variables you created that holds the information you want to log.

A *named list* is simply a list that has names for each element in the list. Putting the information in a named list makes it easy to convert to a data frame. The code in this step describes how to build one.

The code you use to do so is as follows:

```
list_insert = list("DATESTAMP" = datestamp,
                   "TABLENAME" = table_name,
                   "NUM_RECORDS" = num_records
        )
```

The preceding code builds a *named list* that contains the datestamp, table_name, and num_records *scalars* and gives them the names DATESTAMP, TABLENAME, and NUM_RECORDS, respectively.

A scalar is a vector with a length one.

Next, you convert the named list to a data frame using the following code:

```
df_insert = as_data_frame(list_insert)
```

The preceding code converts the named list into a data frame using the *as_data_frame()* function and assigns the results to the df_insert data frame. The names in the named list are used as the column headers.

Step 11: Insert the information you gathered in Step 10 into the history log table

Now that the data you want to log is in a data frame, you can insert it into the *LoadHistoryLog* table using the *sqlSave()* function from the *RODBC* package. You do so using the following code:

```
sqlSave(
        channel = conn,
        dat = df_insert,
        tablename = "LoadHistoryLog",
        append = TRUE,
        rownames=FALSE
)
```

The channel argument in the *sqlSave()* function uses the connection that we created earlier. It inserts the *df_insert* data frame into the *LoadHistoryLog* table. It is very important that you set the *append* argument to TRUE so that *sqlSave()* knows to insert the data into a table that already exists. If it is set to FALSE, it will produce an error because it will try to create a table that already exists. Also, it is important to set the *rownames* to FALSE, because if you don't, a column will be created that has the row names of the data frame and that information is not needed.

Step 12: Close your connection

Next, you need to close your connection using the *odbcClose()* function from the RODBC package. You do so using the following code:

```
odbcClose(conn)
```

Step 13: Copy the script into Power BI

Here is the complete script:

```
library(tidyverse)
library(RODBC)
library(lubridate)

table_name <- "DimDate"
sql <- paste0(
    "SELECT * FROM dbo.",
    table_name)

conn <- odbcConnect(dsn = "SQLServer2017")
df_read <- sqlQuery(channel = conn, query = sql)

datestamp <- now()
num_records <- nrow(df_read)

list_insert <- list(
    "DATESTAMP" = datestamp,
    "TABLENAME" = table_name,
    "NUM_RECORDS" = num_records)

df_insert <- as_data_frame(list_insert)

sqlSave(
    channel = conn,
    dat = df_insert,
    tablename = "LoadHistoryLog",
    append = TRUE,
    rownames=FALSE)

odbcClose(conn)
```

As in previous examples in the book, copy the R code into Power BI via *GetData*. Two data frames, *df_read* and *df_insert*, are produced in the script so you need to tell Power BI which one you want to load into the Power BI data model. You want to load *df_read,* so select it, then click the *Edit* button. Now you need to change the name of the query. Rename the query from *df_read* to *DimDate* in the *PROPERTIES* section. Click *Close & Apply* to load the data into the Power BI data model.

Step 14: Create a script to load DimProduct based on the ReadLog_DimDate.R script

You need to reproduce the preceding steps for four more tables. You can easily reproduce the steps for each table by making a copy of the *ReadLog_DimDate.R* script, then make one minor change to the copied scripts. The change you need to make is the value that the *tablename* variable is set to. In this step, you will do it for the *DimProduct* table. Create a copy of the *ReadLog_DimDate.R* script and change the value of the *table_name* variable from *DimDate* to *DimProduct*. Next, perform Step 13 with the exception being that you rename the query to *DimProduct* instead. Here is what the code looks like after you make the modifications:

```
library(tidyverse)
library(RODBC)
library(lubridate)

table_name <- "DimProduct"
sql <- paste0(
    "SELECT * FROM dbo.",
    table_name)

conn <- odbcConnect(dsn = "SQLServer2017")
df_read <- sqlQuery(channel = conn, query = sql)

datestamp <- now()
num_records <- nrow(df_read)

list_insert <- list(
    "DATESTAMP" = datestamp,
    "TABLENAME" = table_name,
    "NUM_RECORDS" = num_records)
```

```
df_insert <- as_data_frame(list_insert)

sqlSave(
    channel = conn,
    dat = df_insert,
    tablename = "LoadHistoryLog",
    append = TRUE,
    rownames=FALSE)

odbcClose(conn)
```

Step 15: Create a script to load DimPromotion based on the ReadLog_DimDate.R script

Create a copy of the *ReadLog_DimDate.R* script and change the value of the *table_name* variable from *DimDate* to *DimPromotion*. Next, perform Step 13 with the exception being that you rename the query from *df_read* to *DimPromotion*. Here is what the code looks like after you make the modifications:

```
library(tidyverse)
library(RODBC)
library(lubridate)

table_name <- "DimPromotion"

sql <- paste0(
    "SELECT * FROM dbo.",
    table_name)

conn <- odbcConnect(dsn = "SQLServer2017")
df_read <- sqlQuery(channel = conn, query = sql)

datestamp <- now()
num_records <- nrow(df_read)

list_insert <- list(
    "DATESTAMP" = datestamp,
    "TABLENAME" = table_name,
    "NUM_RECORDS" = num_records)
```

```
df_insert <- as_data_frame(list_insert)

sqlSave(
    channel = conn,
    dat = df_insert,
    tablename = "LoadHistoryLog",
    append = TRUE,
    rownames=FALSE)

odbcClose(conn)
```

Step 16: Create a script to load DimSalesTerritory based on the ReadLog_DimDate.R script

Create a copy of the *ReadLog_DimDate.R* script and change the value of the *table_name* variable from *DimDate* to *DimSalesTerritory*. Next, perform Step 13 with the exception being that you rename the query from *df_read* to *DimSalesTerritory*. Here is what the code looks like after you make the modifications:

```
library(tidyverse)
library(RODBC)
library(lubridate)

table_name <- "DimSalesTerritory"
sql <- paste0(
    "SELECT * FROM dbo.",
    table_name)

conn <- odbcConnect(dsn = "SQLServer2017")
df_read <- sqlQuery(channel = conn, query = sql)

datestamp <- now()
num_records <- nrow(df_read)

list_insert <- list(
    "DATESTAMP" = datestamp,
    "TABLENAME" = table_name,
    "NUM_RECORDS" = num_records)
```

```
df_insert <- as_data_frame(list_insert)

sqlSave(
    channel = conn,
    dat = df_insert,
    tablename = "LoadHistoryLog",
    append = TRUE,
    rownames=FALSE)

odbcClose(conn)
```

Step 17: Create a script to load FactInternetSales based on the ReadLog_DimDate.R script

Create a copy of the *ReadLog_DimDate.R* script and change the value of the *table_name* variable from *DimDate* to *FactInternetSales*. Next, perform Step 13 with the exception being that you rename the query from *df_read* to *FactInternetSales*. Here is what the code looks like after you make the modifications:

```
library(tidyverse)
library(RODBC)
library(lubridate)

table_name <- "FactInternetSales"
sql <- paste0(
    "SELECT * FROM dbo.",
    table_name)

conn <- odbcConnect(dsn = "SQLServer2017")
df_read <- sqlQuery(channel = conn, query = sql)

datestamp <- now()
num_records <- nrow(df_read)

list_insert <- list(
    "DATESTAMP" = datestamp,
    "TABLENAME" = table_name,
    "NUM_RECORDS" = num_records)
```

```
df_insert <- as_data_frame(list_insert)

sqlSave(
    channel = conn,
    dat = df_insert,
    tablename = "LoadHistoryLog",
    append = TRUE,
    rownames=FALSE)

odbcClose(conn)
```

Performing the preceding steps not only results in the tables being loaded into Power BI, but also the number of records that were loaded from each table will be recorded in the log table.

Reading SQL Server data using Python

In this section, you will perform the same task that you just performed in R using Python. The logic is similar, but you will be able to perform the task in fewer steps. Let's start with Step 1.

Step 1: Create a DSN to the SQL Server database

If you did the R example, then you already created the *DSN* so you can proceed to the next step. If you did not, you must complete step 1 in the R example before proceeding to the next step. The *DSN* is independent of R and Python so you can use the same one for both.

Step 2: Create a log table in SQL Server

A log table with the name *dbo.LoadHistoryLog* should already be in the database. Execute the following code in *SQL Server Management Studio (SSMS)* if you need to create the *dbo.LoadHistoryLog* table:

```
USE AdventureWorksDW_StarSchema

CREATE TABLE dbo.[LoadHistoryLog] (
    DATESTAMP DATETIME,
    TABLENAME VARCHAR(25),
    NUM_RECORDS INT
)
```

To execute the code in SSMS, you

1. Copy the preceding code

2. Expand the *Database* folder in SSMS, then select *AdventureWorksDW_StarSchema* database

3. Click the *New Query* button located in the menu bar

4. Paste the code you copied in your clipboard to the editor, then click the *Execute* button in the menu bar to execute the code

Step 3: Begin creating the script to load the DimDate table

You will use three libraries in this script. The libraries you will use are *pandas*, *sqlalchemy*, and the *datetime. Pandas* will be the workhorse package as it will be used to read the data that you want to load to the Power BI data model and to write the information back to SQL Server that you want to log. The *create_engine()* function from the *sqlalchemy* library is used to create a connection object to the SQL Server database, and the *datetime* module from the *datetime* library will be leveraged for its date functions. The following code is used to load those libraries, functions, and modules mentioned earlier:

```
import pandas as pd
from sqlalchemy import create_engine
from datetime import datetime
```

Step 4: Create a variable that holds the name of the table that you want to read into Power BI

The name of the table that will be read into the Power BI data model will be used in several places in the script. A variable is used instead of hard-coding the table name. This makes it easier to maintain the script and reuse it for other tables. Here is the code that is used to create the variable that will hold the table name:

```
tablename = "DimDate"
```

Step 5: Create a connection to the database using sqlalchemy

You create a connection to the SQL Server database using the following code:

```
con = create_engine("mssql+pyodbc://SQLServer2017")
```

Just as in the R example, using a DSN makes it very easy to connect to the database. Here you use the *create_engine()* function from the *sqlalchemy* library to make the connection. The syntax used to build the connection string for SQL Server in *sqlalchemy* is `"mssql+pyodbc://<DSN>"`, so in this situation, it is `"mssql+pyodbc://SQLServer2017"`.

Step 6: Read in the contents of the DimDate table and store it as a data frame in the df_read variable

You will use the *read_sql_table()* method from pandas to read the contents of the *DimDate* table and store it in a data frame named *df_read*. The *read_sql_table()* method uses two arguments, the name of the sql table you want to read and the connection to the database. Here is the code:

```
df_read = pd.read_sql_table(tablename, conn)
```

Step 7: Get the current date and time and store the information into the datastamp variable

You perform this task using the following code:

```
datestamp = datetime.now().strftime('%Y-%m-%d %H:%M:%S')
```

You use the *now()* method from *datetime* module to get the time of the load. Unfortunately, the now function returns the date in a format that is not usable by SQL Server so it needs to be reformatted. The format that the date is returned in is *datetime. datetime(YYYY, M, D, H, M, SS, MS)*. So, as an example, January 1, 2019 10:00 PM would be returned as *datetime.datetime(2019, 1, 1, 22, 0, 0, 0)*. You reformat that output using the *strftime()* method to put it in a *YYYY-MM-DD HH:MM:SS* format by applying the following format string: '%Y-%m-%d %H:%M:%S'. That action reformats the *datetime. datetime(2019, 1, 1, 22, 0, 0, 0)* to *'2019-04-14 22:05:24'*. Now you have the datestamp in a format that is usable by SQL Server.

Step 8: Calculate the number of records in the DimDate table

You need to log the number of records that were loaded from the *DimDate* table. To get that information, you use the *shape* method from pandas as illustrated in the following code:

```
num_records = df.shape[0]
```

The shape method returns a tuple with two elements. The first element contains the number of rows in the data frame, and the second element contains the number of columns. You just need the number of rows so you can subset the shape method with [0] to get that information.

A *tuple* is a python data structure that represents a sequence of immutable objects. They are similar to lists with the difference being that they cannot be changed after they are created.

Step 9: Create a one record pandas data frame that contains the information you want to log

You accomplish this in two steps using the following code:

```
dict_insert = {
    "DATESTAMP":[datestamp],
    "TABLENAME":[tablename],
    "NUM_RECORDS":[num_records]}

df_insert = pd.DataFrame.from_dict(dict_insert)
```

The first part of the code creates a *dictionary* of *key/value* pairs. The *key* is a hard-coded value that will be the column name. The value is a *list,* and the elements in the *list* will populate the column. In this case, each *list* only contains one element.

The second part of the code uses the *from_dict* method of the pandas *DataFrame* class to create a data frame. The *dict_insert* dictionary is passed to it, and the result is the *df_insert* data frame.

Step 10: Insert the information you gathered in Step 9 into the history log table

To get the data from the *df_insert* data frame into the *LoadHistoryLog* table, you use the *to_sql* method of the *df_insert* data frame as illustrated here:

```
df_insert.to_sql(
    name='LoadHistoryLog',
    con=conn,
    index=False,
    if_exists = 'append'
)
```

The *name* of the table is *LoadHistoryLog,* the *conn* object you created earlier is used for the *con* argument, the *index* property is set to *False* to prevent pandas from trying to include the index value in the insert, and the *if_exists* is set to *append* so that pandas knows to add the data as an insert if the table already exists.

Step 11: Copy the script into Power BI

Here is the complete script:

```python
import pandas as pd
from sqlalchemy import create_engine
from datetime import datetime

tablename = "DimDate"

conn = create_engine("mssql+pyodbc://SQLServer2017")

df_read = pd.read_sql_table(tablename, conn)

datestamp = datetime.now().strftime('%Y-%m-%d %H:%M:%S')
num_records = df_read.shape[0]

dict_insert = {
    "DATESTAMP":[datestamp],
    "TABLENAME":[tablename],
    "NUM_RECORDS":[num_records]}

df_insert = pd.DataFrame.from_dict(dict_insert)

df_insert.to_sql(
    name='LoadHistoryLog',
    con=conn,
    index=False,
    if_exists = 'append'
)
```

As in previous examples, copy the Python code into Power BI via *GetData*. Two data frames, *df_read* and *df_insert*, are produced in the script, so you need to tell Power BI which one you want to load into the Power BI data model. You want to load *df_read*, so select it, then click the *Edit* button. Next, change the name of the query from *df_read* to *DimDate* in the *PROPERTIES* section. Click *Close & Apply* to load the data into the Power BI data model.

Step 12: Create a script to load DimProduct based on the ReadLog_DimDate.py script

You need to reproduce the preceding steps for four more tables. You can easily reproduce the steps for each table by making a copy of the ReadLog_DimDate.py script, then make one minor change to the copied scripts. The change you need to make is the value that the *tablename* variable is set to. In this step, you will do it for the *DimProduct* table. Create a copy of the *ReadLog_DimDate.py* script and change the value of the *table_name* variable from *DimDate* to *DimProduct*. Next, perform Step 11 with the exception being that you rename the query to *DimProduct* instead. Here is what the script looks like after the changes:

```python
import pandas as pd
from sqlalchemy import create_engine
from datetime import datetime

tablename = "DimProduct"

conn = create_engine("mssql+pyodbc://SQLServer2017")

df_read = pd.read_sql_table(tablename, conn)

datestamp = datetime.now().strftime('%Y-%m-%d %H:%M:%S')
num_records = df_read.shape[0]

dict_insert = {
    "DATESTAMP":[datestamp],
    "TABLENAME":[tablename],
    "NUM_RECORDS":[num_records]}
df_insert = pd.DataFrame.from_dict(dict_insert)

df_insert.to_sql(
    name='LoadHistoryLog',
    con=conn,
    index=False,
    if_exists = 'append'
)
```

Step 13: Create a script to load DimPromotion based on the ReadLog_DimDate.py script

Create a copy of the *ReadLog_DimDate.py* script and change the value of the *table_name* variable from *DimDate* to *DimPromotion*. Next, perform Step 11 with the exception being that you rename the query from *df_read* to *DimPromotion*. Here is what the script looks like after the changes:

```python
import pandas as pd
from sqlalchemy import create_engine
from datetime import datetime

tablename = "DimPromotion"

conn = create_engine("mssql+pyodbc://SQLServer2017")

df_read = pd.read_sql_table(tablename, conn)

datestamp = datetime.now().strftime('%Y-%m-%d %H:%M:%S')
num_records = df_read.shape[0]

dict_insert = {
    "DATESTAMP":[datestamp],
    "TABLENAME":[tablename],
    "NUM_RECORDS":[num_records]}
df_insert = pd.DataFrame.from_dict(dict_insert)

df_insert.to_sql(
    name='LoadHistoryLog',
    con=conn,
    index=False,
    if_exists = 'append'
)
```

Step 14: Create a script to load DimSalesTerritory based on the ReadLog_DimDate.py script

Create a copy of the *ReadLog_DimDate.py* script and change the value of the *table_name* variable from *DimDate* to *DimSalesTerritory*. Next, perform Step 11 with the exception being that you rename the query from *df_read* to *DimSalesTerritory*. Here is what the script looks like after the changes:

```python
import pandas as pd
from sqlalchemy import create_engine
from datetime import datetime

tablename = "DimSalesTerritory"

conn = create_engine("mssql+pyodbc://SQLServer2017")

df_read = pd.read_sql_table(tablename, conn)

datestamp = datetime.now().strftime('%Y-%m-%d %H:%M:%S')
num_records = df_read.shape[0]

dict_insert = {
    "DATESTAMP":[datestamp],
    "TABLENAME":[tablename],
    "NUM_RECORDS":[num_records]}
df_insert = pd.DataFrame.from_dict(dict_insert)

df_insert.to_sql(
    name='LoadHistoryLog',
    con=conn,
    index=False,
    if_exists = 'append'
)
```

Step 15: Create a script to load FactInternetSales based on the ReadLog_DimDate.py script

Create a copy of the *ReadLog_DimDate.py* script and change the value of the *table_name* variable from *DimDate* to *FactInternetSales*. Next, perform Step 11 with the exception being that you rename the query from *df_read* to *FactInternetSales*. Here is what the script looks like after the changes:

```python
import pandas as pd
from sqlalchemy import create_engine
from datetime import datetime

tablename = "FactInternetSales"

conn = create_engine("mssql+pyodbc://SQLServer2017")

df_read = pd.read_sql_table(tablename, conn)

datestamp = datetime.now().strftime('%Y-%m-%d %H:%M:%S')
num_records = df_read.shape[0]

dict_insert = {
    "DATESTAMP":[datestamp],
    "TABLENAME":[tablename],
    "NUM_RECORDS":[num_records]}
df_insert = pd.DataFrame.from_dict(dict_insert)

df_insert.to_sql(
    name='LoadHistoryLog',
    con=conn,
    index=False,
    if_exists = 'append'
)
```

You were able to perform a task that is hard to do in Power Query but relatively easy to do in both R and Python. Power Query can handle reading in data from SQL Server tables but is not able to log information about the load. R and Python have packages and libraries, respectively, that you can leverage to make the task relatively easy to do.

## CHAPTER 6

Reading Data into the Power BI Data Model via an API

Many governmental entities and private companies have made their data more accessible via *Data APIs*. Data APIs provide interfaces that enable you to programmatically retrieve data from data providers. The data sets that these Data APIs provide can bring great value to your Power BI data model.

Interfacing with Data APIs is hard to do using M code, but that is not the case with R and Python. In this chapter, you will learn how to retrieve and load data from the *US Census* via their Data API into the Power BI data model using R and Python. First, you'll see how to load the data using R. Then you'll see the same data loaded again using Python.

Reading Census data into Power BI via an API using R

Here is the scenario. You are a marketing analyst for *Fleek*. Fleek is a retail clothing store whose target market is young adults. The company has a strong presence in the southeastern part of the United States, but they don't have any stores in the state of Indiana. You are tasked with identifying the counties in Indiana that have the demographics that are known to patronize Fleek stores. You want to use demographic data from the US Census to help you with your analysis that you will do in Power BI. The following sections show the steps to perform the task of acquiring demographic data from the US Census using R.

© Ryan Wade 2020
R. Wade, *Advanced Analytics in Power BI with R and Python*, https://doi.org/10.1007/978-1-4842-5829-3_6

Step 1: Get a personal Census API key

To access US Census data, you will need a Census API key. You can obtain one at the following URL: `https://api.census.gov/data/key_signup.html`. You need to fill out a short form and accept the *terms of service*. After you click the *Submit Key Request* button, the US Census Bureau will respond with an email that contains your API key. Make sure you save the key in a safe location.

Step 2: Load the necessary R packages

The packages that you will use in this example are *tidyverse* and *tidycensus*. At this point of the book, you are very familiar with the *tidyverse* package so there is no need for explanation, but that is not the case with the *tidycensus* package. The *tidycensus* package enables you to retrieve data from the US Census and the American Community Survey. The package returns the data in a tidyverse-ready data frame. You can go to the following link to find more information about how to use this package as well as the available data sets: `https://walkerke.github.io/tidycensus/`. Use the following code to load the required libraries and your Census API key that you received from the US Census into your current session using the following code:

```
library(tidycensus)
library(tidyverse)

census_api_key("<census api key>", install = FALSE)
```

If you use the preceding method, you will need to include the code in all future runs. If you want to install the API key in your R environment for future use, then you need to use the following code instead:

```
census_api_key("<census api key>", install = TRUE)
readRenviron("~/.Renviron")
```

Setting the install argument to *TRUE* installs the key in your R environment for future use. The next line of code reloads your environment so that you can use the API key in your current session without having to restart R.

Step 3: Identify the variables you want to return from your data set

The US Census makes available several data sets. The information you need in this example is contained in the *acs5* data set.

The acs5 data set is based on estimates taken over a 5-year period. The other available data sets are *sf1*, *sf3*, *acs1*, *acs3*, *acs5*, *acs1/profile*, *acs3/profile*, *acs5/ profile*, *acs1/subject*, *acs3/subject*, and *acs5/subject*. You can go to `www.census.gov/programs-surveys/acs/guidance/handbooks.html` for more information about the type of data that the US Census offers.

You need population data broken out in age bands from the acs5 data set for calendar year 2015, but you don't know the names of the variables that contain that information. Use the following code to get the entire list of variables that are available in the acs5 data set for calendar year 2015:

```
v15 <- load_variables(year = 2015, dataset= "acs5", cache = FALSE)
View(v15)
```

Let's take a moment to explain the code. The *load_variables* function creates a data frame that warehouses the variables based on the values of the arguments passed to it. The cache option gives you the ability to locally cache the data for faster subsequent retrievals. The output of the load_variables function is depicted in the following table:

name	label	concept
1	I B06001_001E	I Estimate!!Total
2	I B06001_002E	I Estimate!!Total!!Under 5 years
3	I B06001_003E	I Estimate!!Total!!5 to 17 years
4	I B06001_004E	I Estimate!!Total!!18 to 24 years
5	I B06001_005E	I Estimate!!Total!!25 to 34 years

Only the five records of the data set are listed in the preceding output. The resulting data set actually contains a list of over 22,000 variables! Using the View function enables you to view the list of variables in a tabular structure with filtering capabilities. Even with the filtering capabilities, it can be challenging finding the variables you need. In future chapters, you will learn more advanced filtering techniques that will help you find the variables you need via the dplyr package. I used some of those techniques to find the variables that are used in the next step.

Step 4: Create a character vector of the tables that contains the variables you want to return

After you identified the variables you want, you need to place them in a character vector. Here is a character vector of the variables that warehouses the population by age bands that is needed in this example:

```
cns_vars <- c("B06001_001E","B06001_002E","B06001_003E",
"B06001_004E","B06001_005E","B06001_006E","B06001_007E",
"B06001_008E","B06001_009E","B06001_010E","B06001_011E",
"B06001_012E")
```

B06001_001E represents the total population, B06001_002E represents the population under 5, B06001_003E represents the population between 5 and 17, B06001_004E represents the population between 18 and 24, B06001_005E represents the population between 25 and 34, B06001_006E represents the population between 35 and 44, B06001_007E represents the population between 45 and 54, B06001_008E represents the population between 55 and 59, B06001_009E represents the population between 60 and 61, B06001_010E represents the population between 62 and 64, B06001_011E represents the population between 65 and 74, and B06001_012E represents the population that is 75 years of age and above.

Step 5: Configure the get_acs function

Configure the get_acs function to retrieve the data and assign the output to a data frame named *IN_POP_BY_COUNTY_BY_AGE*. The *get_acs()* function retrieves the data from the US Census API and puts it into an R data frame. The following code gets the required population information by age bands at the county level for Indiana:

```
IN_POP_BY_COUNTY_BY_AGE <-
  suppressMessages(
    get_acs(
      geography = "county",
      state = "IN",
      variables = cns_vars,
      survey = "acs5",
      year = 2015,
      output = "wide"
    )
  )
```

Let's explain the code. The *geography* argument tells tidycensus what level of geography you want your data in. The level you want is county, so you use that as the value. To limit the counties to only counties in the state of Indiana, you set the *state* argument equal to "IN". You set the *variables* argument to the character vector that contains the list of variables that you defined in Step 5. The *survey* argument is used to tell tidycensus which data set you want to search. The *year* argument is the year of the survey. The *output* argument tells tidycensus how you want to output the data. Setting it to "wide" puts the data in a format where each variable has its own column. The whole function is wrapped in a *suppressMessages* function. This function suppresses the *get_acs()* function from printing information to the console which can cause problems with Power BI.

Step 6: Give the variables (columns) meaningful names

The names of the columns that are given to us by the US Census are somewhat cryptic, and they do not give you a good idea of what they represent. Fortunately, you were given what they represented when you used the load_variables function to find them. You can rename the columns with more meaningful names using the *rename* function from dplyr. Here is the code that performs that task:

```
IN_POP_BY_COUNTY_BY_AGE =
  rename(
    IN_POP_BY_COUNTY_BY_AGE,
    `Total`=B06001_001E,
    `Under 5`=B06001_002E,
```

```
    `5 to 17`=B06001_003E,
    `18 to 24`=B06001_004E,
    `25 to 34`=B06001_005E,
    `35 to 44`=B06001_006E,
    `45 to 54`=B06001_007E,
    `55 to 59`=B06001_008E,
    `60 and 61`=B06001_009E,
    `62 to 64`=B06001_010E,
    `65 to 74`=B06001_011E,
    `75+`=B06001_012E
  )
```

The first argument of the rename function is the data set that contains the columns that you want to rename. That is followed by pairs of values that have the new column name on the left side and the old column name on the right side separated with a "=" sign.

One of the benefits of using the *rename* function from *dplyr* to rename the column is that you are able to use column names that are not acceptable in regular R data frames.

Step 7: Copy the script into Power BI

Here is the complete script:

```
library(tidycensus)
library(tidyverse)

census_api_key(
    "<census api key>", overwrite = FALSE,
    install = FALSE)

cns_vars <- c("B06001_001E","B06001_002E","B06001_003E","B06001_004E","B060
01_005E","B06001_006E","B06001_007E","B06001_008E","B06001_009E","B06001_01
0E","B06001_011E","B06001_012E")

IN_POP_BY_COUNTY_BY_AGE <-
  suppressMessages(
    get_acs(
      geography = "county",
      state = "IN",
```

```
    variables = cns_vars,
    survey = "acs1",
    year = 2015,
    output = "wide"
  )
)
IN_POP_BY_COUNTY_BY_AGE =
  rename(
    IN_POP_BY_COUNTY_BY_AGE,
    `Total`=B06001_001E,
    `Under 5`=B06001_002E,
    `5 to 17`=B06001_003E,
    `18 to 24`=B06001_004E,
    `25 to 34`=B06001_005E,
    `35 to 44`=B06001_006E,
    `45 to 54`=B06001_007E,
    `55 to 59`=B06001_008E,
    `60 and 61`=B06001_009E,
    `62 to 64`=B06001_010E,
    `65 to 74`=B06001_011E,
    `75+`=B06001_012E
  )
```

Copy the entire script into the R script editor in Power BI. The name of the data frame will be *IN_POP_BY_COUNTY_BY_AGE* by default. If you want to rename the query, then you need to go into edit mode to do so. If you are happy with the name, you can load the data set directly into the Power BI data model to use as the source for some awesome visualizations.

Reading Census data into Power BI via an API using Python

You also have the option of using Python to load data. The following section described the steps to bring in the Census data into Power BI using Python. The workflow is similar to the workflow involving R.

Step 1: Get a personal Census API key

The Census API key is optional to access Census data using the censusdata library from Python. To get one, go to the following URL: `https://api.census.gov/data/key_signup.html`. You need to fill out a short form, accept the *terms of service*, then click the *Submit Key Request* button. The Census Bureau will respond with an email that contains your API key. Make sure you save the key in a safe location.

Step 2: Load the necessary Python libraries

The two libraries needed in this script are the *pandas* and *censusdata* library. Pandas is used for data manipulation, and the censusdata library is used to interface with the Census API. Here is the code needed to load the libraries:

```
import pandas as pd
import censusdata
```

Please refer to the book's introduction to get instructions of how to install python libraries if the preceding libraries are not installed.

Step 3: Identify the variables you want to return in your data set

You need to search the *acs5* data set for variables that contain calendar year 2015 population information by age bands. Type the following code in your console to perform the search:

```
v15 = censusdata.search(
    'acs5',2015,'concept',
    'PLACE OF BIRTH BY AGE IN THE UNITED STATES')
```

The preceding code is not part of the main script. It is used to search for the variables that are available based on the preceding criteria. A script is available in the GitHub repo for this chapter that not only retrieves the variable info but also saves the information as a csv file if you prefer to explore the data in that format.

The preceding code uses the search function from the censusdata library. The first argument is the survey that contains the variables you want to use, the second argument is the year you want to search, and the third argument is the field you want to search. You can choose *label* or *concept*. You used the concept because that field contains the description of the variables you are looking for. The last argument is the search criteria.

The *censusdata.search* function returns a list of tuples.

A *tuple* is a sequence of objects that cannot be changed after being set. The big difference between tuples and lists in python is that lists can be changed after the fact but tuples cannot.

Each tuple contains three elements. The first element in each tuple is the variable name, the second element is the concept, and the third element is the label. The list returned in the search using the preceding code contains 120 tuples. You can subset the first five elements of the tuple to get an idea of what the information looks like using the following code:

```
v15[0:5]
```

The preceding code subsets the v15 list by getting the elements starting at position 0 and ending at but not including position 5. The result is a list that includes the elements in positions 0, 1, 2, 3, and 4. A reformatted version of the output is listed as follows. The output was reformatted, so you can easily identify each tuple:

```
[
    ('B06001_001E',
    'B06001.  PLACE OF BIRTH BY AGE IN THE UNITED STATES',
    'Total:'),
    ('B06001_001M',
    'B06001.  PLACE OF BIRTH BY AGE IN THE UNITED STATES',
    'Margin Of Error For!!Total:'),
    ('B06001_002E',
    'B06001.  PLACE OF BIRTH BY AGE IN THE UNITED STATES',
    'Under 5 years'),
```

```
('B06001_002M',
'B06001.  PLACE OF BIRTH BY AGE IN THE UNITED STATES',
'Margin Of Error For!!Under 5 years'),
('B06001_003E',
'B06001.  PLACE OF BIRTH BY AGE IN THE UNITED STATES',
'5 to 17 years')
]
```

Step 4: Create a variable that is based on the list variables you want

After you identify the variables you want, you need to put them in a python list. Here is the python list that contains the variables that have the population for each age band that we need:

```
cns_vars = [
    "B06001_001E","B06001_002E","B06001_003E","B06001_004E",
    "B06001_005E","B06001_006E","B06001_007E","B06001_008E",
    "B06001_009E","B06001_010E","B06001_011E","B06001_012E"
    ]
```

B06001_001E represents the total population, B06001_002E represents the population under 5, B06001_003E represents the population between 5 and 17, B06001_004E represents the population between 18 and 24, B06001_005E represents the population between 25 and 34, B06001_006E represents the population between 35 and 44, B06001_007E represents the population between 45 and 54, B06001_008E represents the population between 55 and 59, B06001_009E represents the population between 60 and 61, B06001_010E represents the population between 62 and 64, B06001_011E represents the population between 65 and 74, and B06001_012E represents the population that is 75 years of age and above.

Step 5: Create a list of tuples that contains the geographies you want to in your data set

Next, you define your geographical requirements in a list of tuples. Each tuple in the list will have two elements with the first element being the type of geographical entity you want to use and the second element being the FIPS code of the actual geographical entity you want to return.

In this example, you need population information for the state of Indiana broken out by county. You define your requirements using the following code:

```
geographies = [('state', '18'),('county', '*')]
```

The first tuple defines the state requirement. You use the value 'state' to identify the tuple as a "state" tuple, and you use the FIPS code 18 to identify Indiana. This tuple limits the entire search to the state of Indiana. If you don't know the FIPS code for Indiana, you can leverage the *us* package to find that information using the following code:

```
import us
us.states.IN.fips
```

The second tuple is used to define the counties you want to return. You use the value 'county' to tell censusdata that you want to return counties, and you use the asterisk because you want to return all 92 counties in Indiana.

Step 6: Retrieve the data using the *censusdata.download()* function

The following code gets the population data that you want from the US Census and puts the data in a data frame named *IN_POP_BY_COUNTY_BY_AGE*:

```
IN_POP_BY_COUNTY_BY_AGE = censusdata.download(
    'acs5', 2015, censusdata.censusgeo(geographies),
    cns_vars)
```

The first argument is used to define the source you want to use which, in this case, is the acs5 survey. The second argument is used to define the year you want to return. The third argument is used to define the geographical requirements. The list you created in Step 5 that contains the geographical requirements is passed to the *censusdata. censusgeo()* function in this argument. The last argument is the *cns_vars* list. The *cns_vars* is a list of variables you created in Step 4 that identifies the variables you want to download.

Step 7: Reset the index of the data frame created in Step 6

The geography information is contained in the index of the data frame, but that information needs to be in a dedicated column in order for it to be available for Power BI. You use the *reset_index* method of the data frame to put the values of the index in a dedicated column. Here is the code to do that:

```
IN_POP_BY_COUNTY_BY_AGE = IN_POP_BY_COUNTY_BY_AGE.reset_index()
```

The result of running the preceding code is the indexes will be moved to a column named *index*. Now that information will be available for Power BI.

Step 8: Define the new column names

The variables that were identified in Step 4 will be used to name the columns in the returned data set. Those names are somewhat cryptic, and they don't give a good representation of what they represent. The pandas package gives you the ability to rename the columns with more meaningful names. The following python dictionary will give pandas the renaming requirements:

```
new_names = {
    'index':'Geography','B06001_001E':'Total',
    'B06001_002E':'Under 5','B06001_003E':'5 to 17',
    'B06001_004E':'18 to 24','B06001_005E':'25 to 34',
    'B06001_006E':'35 to 44','B06001_007E':'45 to 54',
```

```
'B06001_008E':'55 to 59','B06001_009E':'60 and 61',
'B06001_010E':'62 to 64','B06001_011E':'65 to 74',
'B06001_012E':'75+'
}
```

The preceding data is a python dictionary that has key/value pairs that will be used to rename the columns. The *keys* in the key/value pairs represents the old column name, and the *values* in the key/value pairs represents the new column name.

Step 9: Rename columns

The columns are renamed using the following code:

```
IN_POP_BY_COUNTY_BY_AGE = IN_POP_BY_COUNTY_BY_AGE.rename(
    columns=new_names)
```

The rename method of the *IN_POP_BY_COUNTY_BY_AGE* data frame is used to rename the columns. The columns argument uses the new_names dictionary to get the information it needs to rename the columns.

Step 10: Copy the script into Power BI

Here's the entire script:

```
import pandas as pd
import censusdata

cns_vars = [
    "B06001_001E","B06001_002E","B06001_003E","B06001_004E",
    "B06001_005E","B06001_006E","B06001_007E","B06001_008E",
    "B06001_009E","B06001_010E","B06001_011E","B06001_012E"
    ]

geographies = [('state', '18'),('county', '*')]
```

```
IN_POP_BY_COUNTY_BY_AGE = censusdata.download(
    'acs5', 2015, censusdata.censusgeo(geographies),
    cns_vars)

IN_POP_BY_COUNTY_BY_AGE = IN_POP_BY_COUNTY_BY_AGE.reset_index()

new_names = {
    'index':'Geography','B06001_001E':'Total',
    'B06001_002E':'Under 5','B06001_003E':'5 to 17',
    'B06001_004E':'18 to 24','B06001_005E':'25 to 34',
    'B06001_006E':'35 to 44','B06001_007E':'45 to 54',
    'B06001_008E':'55 to 59','B06001_009E':'60 and 61',
    'B06001_010E':'62 to 64','B06001_011E':'65 to 74',
    'B06001_012E':'75+'
    }

IN_POP_BY_COUNTY_BY_AGE = IN_POP_BY_COUNTY_BY_AGE.rename(
    columns=new_names)
```

Copy the entire script into the Python script editor in Power BI. The name of the data frame returned will be *IN_POP_BY_COUNTY_BY_AGE* by default. If you want to change the name, you need to go into edit mode of Power Query to do so. Otherwise, if you are happy with the name, you can load the data set directly into the Power BI data model.

Summary

In this chapter, you learned how to bring in data from the US Census directly into the Power BI data model using R and Python. Bringing disparate data from outside sources like the US Census can greatly enhance your data models. You are not limited to the US Census API as there are many Data APIs that give you programmatic access to some very rich data sources. Once added to Power BI, these data sources can be the basis for some amazing visualizations.

PART III

Transforming Data Using R and Python

CHAPTER 7

Advanced String Manipulation and Pattern Matching

Basic string manipulation and pattern matching tasks can be handled using one of the numerous string functions pre-packaged in Power Query. You can access many of these functions from the graphical interface, and they are also available for use in custom columns. These functions work great for simple situations but are not sufficient in more advanced scenarios. Fortunately, R and Python have advanced tools that enable you to perform advanced string manipulation and pattern matching tasks beyond the capabilities of Power Query.

In this section of the book, you will learn how to leverage R and Python for advanced string manipulation and pattern matching. The focus will be on the use of *regular expression*. Regular expressions, sometimes referred to as regex, are text strings used to describe a search pattern. They are extremely powerful, and they enable you to perform string manipulation and pattern matching that requires advanced business logic that is not possible using the basic matching methods used in Power Query.

In this chapter, you will learn how to

- Mask sensitive data

- Count the number of words and sentences in a memo field

- Scrub names that are not in a valid format

- Test to see if a pattern exists in a string

You will learn these methods using scenarios that you may encounter in business. Let's first learn how to mask sensitive data.

© Ryan Wade 2020
R. Wade, *Advanced Analytics in Power BI with R and Python*, https://doi.org/10.1007/978-1-4842-5829-3_7

Masking sensitive data

Many companies collect personal data on their customers, which makes both the company and their customers vulnerable to security breaches. Most companies have strict guidelines in place to minimize this risk. But sometimes companies need to share data with outside vendors in order to gain more insights about their data. When they do this, they need to make sure that any sensitive data is masked to protect their customers.

In this scenario, the VP of Collections, Marc Lovejoy, is working with the VP of IT, Didi Costas, to develop an analytical reporting system using Power BI. One of the requirements of the analytical system is that they need to perform sentiment analysis on the comments that the collectors create after every call with the debtors. Some of the comments contain SSNs and phone numbers. They want to use *Microsoft Cognitive Services,* but they need to mask the SSNs and telephone numbers before they send the comments to Microsoft Cognitive Services for sentiment analysis.

Microsoft Cognitive Services are a set of machine learning algorithms that were built by Microsoft to help solve certain problems that require AI. You will learn how to perform sentiment analysis in Power BI in a subsequent chapter.

They shared the requirement with their analyst, Tank Wade, and he came up with two solutions: one using R and one using Python. Here is the solution using R.

Masking sensitive data in Power BI using R

We will illustrate in the steps that follows how we can leverage regular expressions to mask data in Power BI via R. Doing this task using native Power Query features would be extremely hard to do if not impossible. Let's go over the required steps!

Step 1: Import tidyverse and stringr

As always, you start off with the package loads. The first package you will load is *tidyverse*. You will use the *dplyr* package from *tidyverse* to add a new field to your data set and the *readr* package from *tidyverse* to read in the data into the data frame. The *stringr* package will do the heavy lifting in this script. You will leverage regular expressions with the *str_replace* function to mask the sensitive data. Here is the code needed to load the required packages:

```
library(tidyverse)
library(stringr)
```

Step 2: Create the scrub data function

One of the nice features about using R is how easy it is to build a custom function that you can use later in your script. The custom function in this script will take in the text, search for SSNs and phone numbers using a regular expression, mask any SSNs and phone numbers it finds, and return the mask data. Here are the steps needed to create the function:

1. Create the shell of the function as shown here:

    ```
    mask_text <- function(unmask_text){
    }
    ```

    The name of the function will be *mask_text*. The function takes one argument named *unmask_text*.

2. Create a regular expression that matches the pattern of a US phone number and assign the value to a variable named phone_pattern:

    ```
    phone_pattern = "([2-9][0-9]{2})[- .]([0-9]{3})[- .]([0-9]{4})"
    ```

    The preceding regular expression is a common expression used to match phone numbers. Regular expressions like these can be easily found via a bing search.

3. Create a regular expression that matches the pattern for a SSN and assign the value to a variable named *ssn_pattern*:

    ```
    ssn_pattern = "\\d{3}-\\d{2}-\\d{4}"
    ```

    As with the phone number example, chances are you can find regular expressions that do a good job of matching a valid SSN format via an Internet search. But for this example, I wanted to show how easy it is to develop a regular expression to match a SSN pattern.

You can identify numbers in regular expression with the \d character combination. Because \ is a special character in regular expression, you need to escape the backlash in \d by adding another \.

Characters such as the \ and . have special meanings in regular expressions. These characters are known as *metacharacters*. If you want to use the literal representation of these characters in a regular expression, you need to escape them with a *backslash* (\). So, if you wanted to search for a *period (.)* in a string, you need to preface it with a \ to tell the regular expression that you want to literally search for a period. Since the \ is itself a metacharacter, you also need to escape it. With that being said, if you want to send \. to the regular expression engine, you need to send the string "\\.".

Now that you know of a way to identify numbers in a regular expression, let's build the basic pattern of a SSN. The basic pattern of a SSN is three digits, followed by a dash, followed by two digits, followed by another dash, and ending with four digits. The regular expression pattern that is held in the *ssn_pattern* mimics that pattern by using a combination of the \\d and {}. The \\d tells the regular expression that you want to match a number, and the number in the {} tells the regular expression the number of occurrences of a digit you want to match. The dashes in the regular expression are placed in the location where you would expect to find them in a SSN number.

4.   The phone numbers are masked using the following code:

```
cleanned_text <- str_replace(
    unmask_text, pattern = phone_pattern,
    replacement = "XXX-XXX-XXXX")
```

In the preceding code, the *str_replace* function is used to mask any phone numbers that are found in the text. What makes this function different than the *Text.Replace* function in Power Query is that you are able to use regular expressions in str_replace, but you can't use them in *Text.Replace* function. You can do very basic pattern matching in Power Query but not the type you can do with regular expressions. The preceding *str_replace* function takes the string that is passed to it and searches it to see if it matches the pattern of the regular expression passed to it. If it finds the string, it replaces it with a mask defined in the *replacement* argument. So, if the string "Ryan's phone number is 555-852-6301 and his SSN is 123-45-6789" is passed to the function, "Ryan's phone number is XXX-XXX-XXXX and his SSN is 123-45-6789" will be returned.

5. The SSNs are masked in the comments using the following code:

```
cleanned_text <- str_replace(
    cleanned_text, pattern = ssn_pattern,
    replacement = "XXX-XX-XXXX")
```

The preceding code takes the results of the phone mask and applies a similar technique to mask SSN using the ssn_pattern. So, if "Ryan's phone number is XXX-XXX-XXXX and his SSN is 123-45-6789" was passed to the function, "Ryan's phone number is XXX-XXX-XXXX and his SSN is XXX-XX-XXXX" will be returned.

6. The last line of the function returns the result of the SSN and phone masking using the *return* function.

Here is the complete code for the function:

```
mask_text <- function(unmask_text){
  phone_pattern = "([2-9][0-9]{2})[- .]([0-9]{3})[- .]([0-9]{4})"
  ssn_pattern = "\\d{3}-\\d{2}-\\d{4}"

  cleanned_text <- str_replace(
    unmask_text, pattern = phone_pattern,
    replacement = "XXX-XXX-XXXX")

  cleanned_text <- str_replace(
    cleanned_text, pattern = ssn_pattern,
    replacement = "XXX-XX-XXXX")

  return(cleanned_text)
}
```

The logic needed to perform the SSN and Phone Number masking is abstracted in the *mask_text* function. So, when we need to perform a SSN mask and Phone Number mask, we can do so with a simple function call. Abstracting the business logic in a function makes the code much more readable and makes it easier to reuse complex business logic without having to rewrite the code.

Step 3: Read the comments into a data frame

Read the contents of the *Comments.csv* file into a data frame and put it in the *df* variable using the following code:

```
df <- read_csv("Comments.csv")
```

If you are curious to see what the data that you read in look like, you can type the following code in the console to return a view that displays the data frame as a formatted tabular data set:

View(df)

Step 4: Mask the phone numbers and ssn numbers in the comment field

In this step, you will apply the *mask_text* function to the *Comment* field. We do so via the *mutate()* function from *dplyr,* which enables us to do an inline modification of the *Comment* field. Here is the code that performs this task:

```
df <-
  df %>%
  mutate(Comment = mask_text(Comment))
```

Step 5: Copy the script into Power BI

In the last step, you copy the entire script and paste it in the R script editor in *GetData.* Here is the complete script:

```
library(tidyverse)
library(stringr)
```

```r
setwd("<path where the Comments.csv file is located>")

mask_text <- function(unmask_text){
  phone_pattern = "([2-9][0-9]{2})[- .]([0-9]{3})[- .]([0-9]{4})"
  ssn_pattern = "\\d{3}-\\d{2}-\\d{4}"

  cleanned_text <- str_replace(
    unmask_text, pattern = phone_pattern,
    replacement = "XXX-XXX-XXXX")

  cleanned_text <- str_replace(
    cleanned_text, pattern = ssn_pattern,
    replacement = "XXX-XX-XXXX")

  return(cleanned_text)
}

df <- read_csv("Comments.csv")

df <-
  df %>%
  mutate(Comment = mask_text(Comment))
```

The resulting data frame with the masked data will be exposed to Power BI as a data set that you can add to Power BI data model. This information can be used to enhance your Power BI data model, which will enable you to create richer visualizations.

Masking sensitive data in Power BI using Python

We used R in the previous example to mask the SSNs and Phone Numbers in the comments. We can do the same with Python using similar logic but just a different programming language. You will find many instances where you can perform a task equally as well in R or Python. The language that you used is often a matter of personal preference. Let's now go over how to perform the masking we just did in R using Python.

Step 1: Import pandas, os, and re library

As always, you start off with the library loads. The first library you will load is pandas. You will use it to read in the data and to mask the data in the *Comment* field. You will also load the os package for some minor file system task, and the re package will be used to handle the regular expression needs in this script. Here is the code that loads the necessary libraries:

```
import pandas as pd
import re
import os
```

Step 2: Create the mask_text function

Next, you need to build a custom function to mask the phone numbers and SSNs that are in the *Comment* field. The function in this script will take in the text from the *Comment* field, find the SSNs and phone numbers in the *Comment* field based on a regular expression pattern match, mask the SSNs and phone numbers, then return the text with the mask data. Here are the steps to create the function:

1. Create the shell of the function as shown here:

    ```
    def mask_text (unmask_text):
    ```

    The name of the function will be mask_text. The function takes one argument named unmask_text.

2. Create a regular expression that matches the pattern of a US phone number and assign the value to a variable named phone_pattern:

    ```
    phone_pattern = r"([2-9][0-9]{2})[- .]([0-9]{3})[- .]([0-9]{4})"
    ```

    The preceding regular expression is a common expression used to match phone numbers. Many common patterns such as patterns to match phone numbers and SSNs are easily found via a Bing or Google searches.

Please note the use of *r* before the regular expression. In Python, but not in versions of R prior to version 4.0.0, you can prefix a string with *r* to tell Python you want to treat the string as a raw string. When you do so, the backslash is treated as its literal representation and not as a special metacharacter. This prevents you from having to escape the backslashes.

3. Create a regular expression that matches the pattern of a SSN and assign the value to a variable named *ssn_patterns*:

```
ssn_pattern = r"\d{3}-\d{2}-\d{4}"
```

The pattern of a SSN is three digits, followed by a dash, followed by two digits, followed by another dash, and ending with four digits. In the preceding regular expression, you use \d to identify a digit and you use *{n}* to tell the regular expression how many times you want to match the preceding pattern. So \d{3} in the regular expression says that you want to match three consecutive digits.

The regular expression in this example is lax and will result in many false-positive matches because only certain digit combinations can be used in a SSN. There are more complicated regular expressions that you can use to reduce the number of false positives, but the higher level of complexity may make them much harder to read. When using regular expressions, you must consider the trade-off of accuracy and interpretability.

4. The phone numbers are masked using the following code:

```
cleanned_text = re.sub(
    phone_pattern, "XXX-XXX-XXXX", unmask_text)
```

In the preceding code, the *re.sub* function is used to mask any phone numbers that are found in the text with Xs. One of the things that make this function different than the *Text.Replace* function in Power Query is that you are able to use regular expressions with this function.

5. The SSNs are masked in the comments using the following code:

```
cleanned_text = re.sub(
    ssn_pattern, "XXX-XX-XXXX", cleanned_text)
```

The code above takes the results of the phone mask and applies a similar technique using the ssn_pattern pattern.

6. The last line of the function returns results of the SSN and phone number masking using the *return* statement.

Here is the complete code needed to create the function:

```
def mask_text (unmask_text):
    phone_pattern = \
        r"([2-9][0-9]{2})[- .]([0-9]{3})[- .]([0-9]{4})"
    ssn_pattern = r"\d{3}-\d{2}-\d{4}"
    cleanned_text = re.sub(
        phone_pattern, "XXX-XXX-XXXX", unmask_text)
    cleanned_text = re.sub(
        ssn_pattern, "XXX-XX-XXXX", cleanned_text)
    return cleanned_text
```

The logic needed to perform the SSN and Phone Number masking is abstracted in the *mask_text* function. So, when you need to mask SSNs and Phone Numbers, you can do so with a simple function call. Abstracting the business logic in a function makes the code much more readable and makes the business logic easier to reuse.

Step 3: Set the working directory

Set the working directory to the location of the *Comments.csv* file using the following code:

```
os.chdir("<path to folder that contains the Comments.csv file")
```

Step 4: Read the comments into a data frame

Read the contents of the *Comments.csv* file into a data frame using *pandas* and assign the data frame to a variable named *df* using the following code:

```
df = pd.read_csv("Comments.csv")
```

Step 5: Mask the phone numbers and SSNs

The following code does an inline update of the *Comment* field by applying *mask_text()* function to the *Comment* field via the *apply()* method of that field:

```
df.Comment = df.Comment.apply(mask_text)
```

Step 6: Copy the script into Power BI

In the last step, you copy the entire script and paste it in the Python script editor in *GetData*. Here is the complete script:

```
import pandas as pd
import re
import os

def mask_text (unmask_text):
    phone_pattern = \
        r"([2-9][0-9]{2})[- .]([0-9]{3})[- .]([0-9]{4})"
    ssn_pattern = r"\d{3}-\d{2}-\d{4}"
    cleanned_text = re.sub(
        phone_pattern, "XXX-XXX-XXXX", unmask_text)
    cleanned_text = re.sub(
        ssn_pattern, "XXX-XX-XXXX", cleanned_text)
    return cleanned_text

os.chdir("<path to folder that contains the Comments.csv file")
df = pd.read_csv("Comments.csv")
df.Comment = df.Comment.apply(mask_text)
df
```

The resulting data frame with the masked data will be exposed to Power BI as a data set that you can add to Power BI model. The added data can be used to help you tell a better data story.

Counting the number of words and sentences in reviews

When many think of data science, they think of algorithms that predict a numerical outcome or the probability of an event occurring. Little do they know you are not limited to doing those types of predictions in data science. A subset of data science includes techniques to work with unstructured text. This area of data science is called Natural Language Processing (NLP). Techniques used in NLP, such as sentiment analysis or topic analysis, are pretty advanced and are hard to build. In the next chapter, we will leverage Microsoft Cognitive Services to perform those types of tasks in Power BI. But there are more simple NLP techniques we can use without the need to make API calls to services like Microsoft Cognitive Services. We will cover two of those techniques in this chapter. The two techniques we will cover are counting the number of words and sentences in a corpus of text. Let's illustrate with an example.

Sarah is a data analyst for Yelp. She has been tasked to add additional information to a report that Yelp gives their customers. The report contains the customer reviews, and Sarah's boss, Jessica, wants her to add a column that contains the number of words in each review and a column that contains the number of sentences in each review. She was able to accomplish both requirements in R but was only able to perform a word count in Python. Here are the steps she used to perform the task in R.

Counting the number of words and sentences in reviews in Power BI using R

In this example, we will use R to calculate both the number of words in a yelp review and the number of sentences in the review. As we have done before, we will lean on the tools in *tidyverse* to help us with this task.

Step 1: Import tidyverse and stringr

The two packages that will be used in this script are the *tidyverse* metapackage and *stringr*. The *dplyr* package from *tidyverse* will be used to add a column to the data set that contains a word count and to add another column that contains the sentence count. The *str_count* function from the *stringr* package is doing the heavy lifting in this script:

```
library(tidyverse)
library(stringr)
```

Step 2: Change working directory to location of the file

As always, you use the setwd() function to set the working directory to the location where your data is. In this example, you set it to the location where the *Yelp* data is using the following code:

```
setwd("<file path to where the file is located>")
```

Step 3: Read in the Yelp data

The *read_csv* function from *readr* is used to read in the Yelp data into an R data frame. The data frame is assigned to the *df* variable using the following code:

```
df <- read_csv("yelp_training_set_review_sample.csv")
```

Step 4: Subset the columns

The Yelp data set is very wide. It contains many columns, and you are only interested in a subset of them. So, you need to create a workflow that starts with the df data frame, then subset it to only include the columns you are interested in. That is done using the following code snippet:

```
df <-
  df %>%
  select(id, business_categories, business_review_count,
         business_stars, business_state, business_type,
         stars,text)
```

Step 5: Add word count and sentence count columns

Now you need to add two columns to your data frame, one named *word_count* and another one named *sent_count*. The *word_count* field will store the number of words in the review, and the *sent_count* field will contain the number of sentences in the review. You will accomplish both tasks using the *str_count* function from *stringr* as illustrated in the following code:

```
mutate(word_count =
    str_count(text, boundary("word"))
    ,sent_count = str_count(text, boundary("sentence")))
)
```

The *str_count* function uses the *boundary* function to specify the type of count you want to do. Notice how easy it is to do a word count and sentence count! It is a simple function call! Performing a sentence count in Power Query is not feasible, but it is a simple function call in R.

Step 6: Copy the script into Power BI

Here is the complete script:

```
library(tidyverse)
library(stringr)

setwd("<path to Yelp file>")

df <- read_csv("yelp_training_set_review_sample.csv")

df <-
  df %>%
  select(
      id
    ,business_categories
    ,business_review_count
    ,business_stars
    ,business_state
    ,business_type,stars,text) %>%
```

```
mutate(
    word_count = str_count(text, boundary("word"))
    ,sent_count = str_count(text, boundary("sentence"))
)
```

Copy the entire script and paste it in the R script editor in *GetData* so you can add the resulting data frame to the Power BI data model. The data set is now available to enhance your Power BI data model to enable you to more effectively tell your data story.

Counting the number of words in reviews in Power BI using Python

In this example, you will use Python to calculate the number of words in a Yelp review. For many tasks, both R and Python will be equally suitable, but there will be situations where one language is more suitable to perform a task than the other. That is the case here. The solution to calculate the number of sentences in a review was not as apparent in Python as it was in R, so we will focus on just doing the word count. The steps you can use to do a word count are listed here.

Step 1: Import pandas and os

The two libraries that you will use in this script are *pandas* and *os*. The *pandas* library will be used to read the data from the *Yelp* file and also to calculate the number of *words* in the review. The *os* package will be used for some file system tasks. Here is the code used to import the two packages:

```
import pandas as pd
import os
```

Step 2: Change working directory to the location of the file

The working directory is set to the location where the Yelp file is located using the *os.chdir()* function. Here is the code needed to perform the task:

```
os.chdir(r"<file path to Yelp file>")
```

You are able to use the traditional file path convention because you prefaced the file path string with *r* to tell python that you want the raw version of the string. A path like C:\Users\Public will work because you will not need to escape the backslashes. This is a feature that has been available in Python for some time now but just came available in R in version 4.0.0.

Step 3: Read the Yelp data into Python

The contents of the Yelp file are read into Python via the *read_csv()* method from *pandas* and assigned to the variable *df* using the following code:

```
df = pd.read_csv("yelp_training_set_review_sample.csv")
```

Step 4: Create the word_count column

This step is accomplished using the following code:

```
df["word_count"] = df["text"].str.split().apply(len)
```

The code is very succinct, but a lot is going on. Let's unpack it. First, pandas uses the *split()* method from the *str* class to split the words in the text and put them in a pandas list. So, if the content in the *text* field was "The cat in the hat.", the split function will split the text and put it in a list that contains five elements that looks like this: [The, cat, in, the, hat.]. The default delimiter for the split function is a white space which is what you need in this situation. If you were to stop here, you will be left with a data frame that has a column named word_count that is populated with a bunch of lists that are based on the text field.

The reason why you converted the text field to a list is to take advantage of a particular list property. You can use the *len* function against a list to get the number of elements in it. In the case here, each element in the list represents a word in the text, so the result of applying the *len* function against it will be a good proxy for the word count. The latter is done by using the *apply()* method to apply the *len* function to the list created by the *split()* function.

Step 5: Copy the script into Power BI

The complete script looks like this:

```
import pandas as pd
import os

os.chdir(r"<file path to the folder where the Yelp file is>")
```

```
df = pd.read_csv("yelp_training_set_review_sample.csv")
df["word_count"] = df["text"].str.split().apply(len)
```

You need to copy this script and paste it into the Python editor in Power BI. The resulting data frame will be available to add to the Power BI data model.

Removing names that are in an invalid format

Data in source systems is often messy and not well formatted. In many instances, the data needs to be scrubbed before it can be used in reporting. That is the case in the example that follows.

In this scenario, Jonathan, a consultant for SQLWeave, is responsible for ETL, and he needs to scrub names from DimEmployees dimension table that are not in a valid format before he loads the data into Power BI. Only names that are in the FN MI LN or FN LN format will be accepted where FN stands for *First Name*, MI stands for *Middle Initial*, and LN stands for *Last Name*. Jonathan developed a script in both R and Python to handle this task. Here are the steps he used to perform the task in R.

Removing names that are in an invalid format in Power BI using R

Here are the steps required to scrub the names using R. We will accomplish the task using tools that are made available in the *tidyverse*. Here are the steps.

Step 1: Import tidyverse and stringr

You will use the *tidyverse* metapackage and *stringr* in this script. The *readr* package from tidyverse is used to read in the data that is contained in the *DimEmployee.csv* file into an R data frame, and *dplyr* from *tidyverse* is used to do an inline modification of the *Name* field. The modification will leverage the *str_extract()* function from the *stringr* package. The following code imports the required packages:

```
library(tidyverse)
library(stringr)
```

Step 2: Change working directory to location where the DimEmployee.csv file is

The *setwd()* function is used to set the working directory to the file path that contains the file:

```
setwd("<file path to DimEmployee.csv file>")
```

Step 3: Create a regular expression to match a valid name

The regular expression that will be used to match a proper name is defined here and assigned to the *nameWithOrWithoutMiddleInitial* variable:

```
nameWithOrWithoutMiddleInitial = "(^[A-Z][a-z]{1,10}\\s[a-zA-Z]\\s[A-Z]
[a-z]{1,10}$)|(^[A-Z][a-z]{1,10}\\s[A-Z][a-z]{1,10}$)"
```

The regular expression may look complicated, but it is actually quite simple. Let's break it down into steps:

1. The regular expression is matching two name patterns: one in the format of *FN MI LN* and other in the format of *FN LN* where *FN* is first name, *MI* is middle initial, and *LN* is last name. The two patterns are separated with a pipe, which stands for a logical "OR". The *FN MI LN* pattern in the regular expression is in blue, and the *FN LN* pattern in the regular expression is in red:

   (^[A-Z][a-z]{1,10}\\s[a-zA-Z\\s[A-Z][a-z]{1,10}$)|(^[A-Z][a-z]{1,10}\\s[A-Z][a-z]{1,10}$)

2. Now let's break down the blue section. Recall that the blue section is matching the pattern *FN MN LN*. We will focus on the *FN* part first. The regular expression uses the following code to identify the *FN*: ^[A-Z][a-z]{1,10}. The ^ tells the regular expression that you are starting the match from the beginning of the string. The ^ is followed by [A-Z] which tells the regular expression that the first character must match a capital letter between A and Z. Next is [a-z]{1,10}. That combination tells the regular expression that the next 1–10 characters should be a lowercase letter between a and z. This gives the regex flexibility to match a first name that has a length of 2–11 letters.

3. The *MI* in the blue part of the regular expression is matched using the following code: \\s[a-zA-Z]\\s. The \\s is used in regular expression to match white spaces. Recall that the \ is a special metacharacter, so if you want to literally use a backslash in your regular expression, you need to escape it with another backslash. The *MI* needs to be a single upper- or lowercase letter, and [a-zA-Z] matches that situation. A white space is needed between the MI and LN, so \\s is used to match that criteria.

4. Last, we need to match the last name on the blue section. That is done using the following pattern: [A-Z][a-z]{1,10}$. The [A-Z] tells the regular expression that the first letter in the *LN* needs to be uppercase. The [a-z]{1,10} combination tells the regular expression that the subsequent letters in the last name must be lowercase and {1,10} says that you can have 1–10 lowercase letters. The $ in the regular expression terminates the match.

5. The *FN LN* match is in red. The only difference is that \\s[a-zA-Z]\\s is removed because you are trying to match a name that does not have a middle initial.

Step 4: Read the data into a data frame

Read in the contents of the *DimEmployee.csv* file into an R data frame and assign it to the df variable using the following code:

```
df <- read_csv("DimEmployee.csv")
```

Step 5: Do an inline update of the *Name* column

The code used to perform this step is listed as follows:

```
df <-
  df %>%
  mutate(Name = str_extract(Name,nameWithOrWithoutMiddleInitial))
```

The *mutate()* function is used to do the inline update. The *str_extract()* function from *stringr* tests the *Name* field to see if it matches the regular expression pattern defined in the *nameWithOrWithoutMiddleInitial* variable. If it finds a match, it will extract the text that matches the pattern from the *Name* field. Otherwise, it returns nothing.

Step 6: Copy the script into Power BI

In the last step, you copy the entire script and paste it in the R script editor in *GetData*. Here is the complete script:

```
library(tidyverse)
library(stringr)

setwd("C:/Users/rwade/Downloads/Chapter5/Chapter5")

nameWithOrWithoutMiddleInitial = "(^[A-Z][a-z]{1,10}\\s[a-zA-Z]\\s[A-Z]
[a-z]{1,10}$)|(^[A-Z][a-z]{1,10}\\s[A-Z][a-z]{1,10}$)"

df <- read_csv("DimEmployee.csv")

df <-
  df %>%
  mutate(Name = str_extract(Name,nameWithOrWithoutMiddleInitial))
```

The resulting data frame with the masked data will be exposed to Power BI as a data set that you can add to Power BI model.

Removing names that are in an invalid format in Power BI using Python

Here are the steps required to scrub the names using Python. We will accomplish the task using *pandas, re,* and *os.* Here are the steps.

Step 1: Import pandas, re, and os library

You will use *pandas, re,* and *os* in this script. The *pandas* library is used to read in the data in the *DimEmployee.csv* file into a Python data frame and is also used to do an inline modification of the *Name* field. The modification will leverage some functions from the *re* library to perform some regular expression tasks. It will also use the *os* package to handle the file system tasks. Here is the code used to import those libraries:

```
import pandas as pd
import re
import os
```

Step 2: Change working directory to the location where the DimEmployee csv file is

The script uses the *os.chdir()* to change the working directory to the location that warehouses the *DimEmployee.csv* file using the following code:

```
os.chdir(r"<file to the DimEmployee.csv file>")
```

Step 3: Read in the DImEmployee data into an R data frame

The *read_csv()* method from the pandas package is used to read in the contents of the DimEmployee.csv file into a pandas data frame, and it is assigned to the *df* variable using the following code:

```
df = pd.read_csv("DimEmployee.csv")
```

Step 4: Create a regular expression that matches a valid name

The regular expression that you will use that matches a proper name is defined as follows:

```
r'([A-Z][a-z]{1,10}\s[a-zA-Z]\s[A-Z][a-z]{1,10})|' \
r'([A-Z][a-z]{1,10}\s[A-Z][a-z]{1,10})'
```

The regular expression is compiled using the compile method from the *re* package, and it is assigned to the *nameWithOrWithoutMiddleInitial* variable. The regular expression may look complicated, but it is actually quite simple. Let's break it down into steps:

1. The regular expression is matching two name patterns: one in the format of *FN MI LN* and other in the format of *FN LN* where *FN* is first name, *MI* is middle initial, and *LN* is last name. The two patterns are separated with a pipe, which stands for a logical "OR". The *FN MI LN* part of the pattern in the regular expression is in blue, and the *FN LN* part of the pattern of the regular expression is in red. The expression is prefaced with *r*, which tells Python to

treat the string as a *raw* string. Raw strings interpret backslashes in a literal fashion and not as a special character, so backslashes in a raw string do not need to be escaped:

r'([A-Z][a-z]{1,10}\s[a-zA-Z\s[A-Z][a-z]{1,10})|' \
r'([A-Z][a-z]{1,10}\s[A-Z][a-z]{1,10})'

2. Now let's break down the blue section. Recall that the blue section is matching the pattern *FN MN LN*. We will focus on the *FN* part first. The regular expression uses the following code to identify the *FN*: [A-Z][a-z]{1,10}. The *[A-Z]* part of the regex says that that the first character in the pattern should match a capital letter between A and Z. Next is [a-z]{1,10}. That combination of characters tells the regular expression that the next 1–10 characters should be a lowercase letter between a and z. This gives the regex flexibility to match a first name that has a length of 2–11 characters.

3. The *MI* in the blue part of the regular expression is matched using the following code: \s[a-zA-Z]\s. The \s is used in regular expression to match a white space. The MI needs to be a single upper- or lowercase letter, and [a-zA-Z] matches that situation. A white space is needed between the *MI* and *LN,* so \s is used to match that criteria. You did not need to use a backslash to escape your backslashes because the regular expression string was converted to a raw string so the backslash is not interpreted as a special character.

4. Next, you need to match the last name. That is done using the following pattern: [A-Z][a-z]{1,10}. The [A-Z] tells the regular expression that the first letter in the *LN* needs to be uppercase. The [a-z]{1,10} combination tells the regular expression that the subsequent letters in the last name must be lowercase, and {1,10} says that you can have 1–10 lowercase letters. The $ in the regular expression terminates the expression.

5. The *FN LN* match is in red. The only difference is that \s[a-zA-Z]\s is removed because we don't need to match a middle initial.

The previous five steps illustrated how simple regular expressions are and how easy it is to match patterns with them. The pattern we used earlier was simple, but there are many features in regular expressions that enable you to match some complicated patterns. They are very powerful and are a must in a data analyst toolkit.

Step 5: Compile the regular expression

The code required to perform this step is given here:

```
namepattern = \
    r'([A-Z][a-z]{1,10}\s[a-zA-Z]\s[A-Z][a-z]{1,10})|' \
    r'([A-Z][a-z]{1,10}\s[A-Z][a-z]{1,10})'

nameWithOrWithoutMiddleInitial = re.compile(namepattern)
```

First, we assign the regular expression that was created in Step 4 to a variable named *namepattern*. Then we create an object pattern named *nameWithOrWithoutMiddleInitial* using the *compile()* method from *re*. The object will be used to perform the pattern matching in later steps.

Step 6: Define a function that executes the name test

The name of the function that performs the scrubbing is *scrubName*. Here is the code:

```
def scrubName(name):
    m = nameWithOrWithoutMiddleInitial.fullmatch(name)
    if m:
        return m.group(0)
    else:
        return None
```

The function takes one argument which is the name you want to scrub. The first line of code in the function uses the *fullmatch()* method from the *nameWithOrWithoutMiddleInitial* regular expression object. This method takes the name argument that was passed to the function and tests to see if the name fully matches the pattern defined in *nameWithOrWithoutMiddleInitial*. Note that you did not have to use a ^ or $ to define the beginning and end of your pattern since you are doing a full match.

The result is stored in *m*. If there is a match, then a *match object* is returned; otherwise, *None* is returned which is Python's equivalent to *null*. The match object has many different methods. The one you will use is the *group()* method. Let's illustrate with a simple regular expression:

```
import re
p = re.compile("(Ryan)\\s(Wade)")
m = p.fullmatch("Ryan Wade")
m.group(0)
m.group(1)
m.group(2)
```

The regular expression used here is a literal expression that matches my name. My first name and last name are put into groups using parentheses. So if I passed the string "Ryan Wade" to *p*, *fullmatch()* will obviously find a match, and it will break the match in three groups. If I want to return the full matched string "Ryan Wade", use *m.group(0)* because 0 returns the full match. If I want to return just "Ryan", use *m.group(1)*, and if I just want to return "Wade", use *m.group(2)*.

So, let's go back to the *scrubName* function. If a match is found, you want to return the entire string. You can retrieve the full match using group 0. If a match is not found, then *None* is returned.

Step 7: Apply the function to the column to scrub the name

An inline update is done using the following code:

```
df["Name"] = df.Name.apply(scrubName)
```

The apply method of the *Name* field is used to apply the scrubName function to the Name field.

Step 8: Copy the script into Power BI

In the last step, you copy the entire script and paste it in the Python script editor in *GetData*. Here is the complete script:

```
import pandas as pd
import re
import os
```

```
os.chdir(r"<file to the DimEmployee.csv file>")
df = pd.read_csv("DimEmployee.csv")

namepattern = \
    r'([A-Z][a-z]{1,10}\s[a-zA-Z]\s[A-Z][a-z]{1,10})|' \
    r'([A-Z][a-z]{1,10}\s[A-Z][a-z]{1,10})'

nameWithOrWithoutMiddleInitial = re.compile(namepattern)

def scrubName(name):
    m = nameWithOrWithoutMiddleInitial.fullmatch(name)
    if m:
        return m.group(0)
    else:
        return None

df["Name"] = df.Name.apply(scrubName)
```

The resulting data frame will be made available to the Power BI data model.

Identifying patterns in strings based on conditional logic

Most string parsing functions can be handled relatively easily using Power Query's native functionality. But when your string parsing is based on conditional logic, the code can get verbose and hard to understand. Fortunately, many situations that require string parsing with conditional logic can be handled more succinctly in R and Python via regular expressions. Let's illustrate how the fictitious company, MC Diesel, solved a business problem by using regular expressions based on conditional business logic to parse a string.

Rodney is the Executive Director for MC Diesel. He knows that MC Diesel had issues in the past with two parts they manufacture. The two parts are fuel pumps (part # 561769) and fuel injectors (part # 561394). The issues were localized to plants in AL, LA, and OH and only occurred in models A and B.

Rodney tasked his data scientist, Annie Su, to develop a dashboard with drill-through capabilities to help him further diagnose and monitor the problem. The only data Annie was able to obtain was a report with information listed in Table 7-1.

Table 7-1. *Parts Report*

Date Key	SKU	Shift	Store ID	Recalled
20190813	75197726D3	3	263	False
20191211	70136522D1	1	221	False
20190401	70195609A3	3	92	False
20191228	75102709B1	1	91	False
20190304	70162016C3	3	161	True
20191218	70136526D2	2	263	False
20190322	70162029D1	1	291	False
20191012	70136539C2	2	395	False
20190318	56176926B2	2	262	False

It appears that the report does not contain any information that identifies warehouse location or part number and that information is needed for the analysis. Annie is very smart, and she remembered that the SKU number is not some arbitrary number. She knows that each part of the number has meaning. She knows that the first six digits represent the part number, the digits in positions 7 and 8 are the state FIPS code of the warehouse where the product was made, the character in position 9 represents the model, and the digit in position 10 represents the shift that the product was made in. So, an item with a SKU of 56176922A2 represents the following:

- A fuel pump because the first six digits, 561769, are the product numbers for fuel pumps

- Was manufactured in Louisiana because the digits in positions 7 and 8, 22, are the state FIPS code for the state of Louisiana

- Has a model type of A because the character in position 9 is "A"

- Was manufactured during the second shift because the digit in position 10 is 2

Annie needs to create a calculated column that returns the *part number* from the SKU if the following conditions are met:

- Is a fuel pumps made in AL, LA, or OH with a model type of A or B.

- Is a fuel injectors made in AL, LA, or OH with a model type of A or B.

If she performed this task using Power Query, the resulting code will be very verbose but much more succinct using R or Python. The following steps illustrate how Annie was able to create a solution using R.

Identifying patterns in strings based on conditional logic in Power BI using R

We will leverage regular expressions to match the products using the criteria defined earlier. This example will illustrate how succinctly we are able to describe the logic needed to find the SKUs we are looking for. The same criteria would be very verbose in Power Query if you had to use M code. Let's go over the steps.

Step 1: Import the tidyverse and stringr packages

The first step is to import the required packages. The *readr* from *tidyverse* is used to read in the contents of the *ProductionOrders.csv* file into an R data frame. The *dplyr* package is used to handle the data frame manipulation tasks, and the *stringr* package is used to handle the string manipulation tasks. Here is the code used to import the preceding packages:

```
library(tidyverse)
library(stringr)
```

Step 2: Change the working directory

The working directory is changed to the file path that contains the *ProductionOrders.csv* file using the *setwd()* function as illustrated in the following code:

```
setwd("<Path to ProductionOrders.csv file>")
```

Step 3: Create a function that identifies the products

The name of the function to handle this task is *monitoredProducts*. Here are the steps to create the function:

1. Create the shell of the function using the following code:

    ```
    monitoredProducts <- function(sku){

    }
    ```

    The *monitoredProducts* product function will require just one parameter, *sku.number*.

2. Next, you need to define a regular expression that finds all fuel pumps (561769) and injectors (561394) that were created in AL, LA, or OH with a model type of A or B. You also want to return just the product number portion of the SKU and not the entire SKU number. You can accomplish this in regular expressions by using a *positive lookahead*.

    The website www.rexegg.com says that a positive lookahead asserts that what immediately follows the current position in the string is the pattern you are looking for. Let's look at the regular expression in this script and break down what that statement means. Here is the regular expression in question:

    ```
    pattern = "^(561769|561394)(?=(01|22|39)(A|B))
    ```

    The positive lookahead portion of the preceding regular expression is (?=(01|22|39)(A|B)). The structure of a positive lookahead is a regular expression grouped in parenthesis that starts with ?=. The pattern that follows the ?= is the pattern that must follow the preceding pattern in order for the match to pass.

    So, the regular expression identified in the *pattern* variable will find a match in the SKU number if it

    -   Starts with 561769 or 561394

    -   Is preceded by state FIPS codes for AL, LA, or OH

    -   Has a model type of A or B

This will make more sense when you use it in the function in Step 5.

3. In this step, you extract out the portion of the SKU that matches your regular expression pattern using the following code:

```
mp = str_extract(sku, pattern = pattern)
```

4. The code uses the *str_extract()* function to extract the strings that match the regular expression you define in Step 2. It will only return the product code portion of the string if there is a match. Let's illustrate how it works with an example. If you passed 56176939A2 to the preceding function, R will return 561769. That is because not only does the string pass the first condition because it starts with 561769 but it also passes the positive lookahead test because positions 7–8 in the string are 39 which is one of the acceptable state FIPS codes and position 9 is A which is one of the acceptable models. Alternatively, 56176910D2 returns NA because the positive lookahead fails. The state FIPS code *10* is not one of the FIPS codes you are looking for. The same hold true for the model because model D is not a model that you are looking for.

Step 4: Read in the data from the ProductionOrders.csv file into an R data frame

The contents of the *ProductionOrders.csv* file are read into R and stored in the *df* variable using the *readr* package using the following code:

```
df <- read_csv("ProductionOrders.csv")
```

Step 5: Add a column to the df named "Monitored Products"

Next, you write code that performs a simple workflow of adding a new column to the data frame named *Monitored Products* using the *mutate()* function from dplyr package. Here is the code needed to perform that action:

```
df <-
  df %>%
  mutate(`Monitored Products` = monitoredProducts(SKU))
```

The preceding code uses the *mutate()* verb from dplyr to create the *Monitored Products* column that is based on the results of applying the *monitoredProducts()* custom function against the SKU column.

Step 6: Copy the script into Power BI

Here is the complete script:

```
library(tidyverse)
library(stringr)

setwd("<Path to ProductionOrders.csv file>")

monitoredProducts <- function(sku){
  pattern = "^(561769|561394)(?=(01|22|39)(A|B))"
  mp = str_match(sku, pattern = pattern)[1]
  return_value = ifelse(is.na(mp),"Not Monitored", mp)
  return(return_value)
}

vmonitoredProducts = Vectorize(monitoredProducts)

df <- read_csv("ProductionOrders.csv")

df <-
  df %>%
  mutate(`Monitored Products` = vmonitoredProducts(SKU))
```

This script creates a data frame named *df* that can be added to the Power BI data model via the *R script* editor in *GetData*.

Identifying patterns in strings based on conditional logic in Power BI using Python

In this section, we will perform the task we just completed in R with Python. As in the R example, the main tool that will be leveraged are regular expressions. You will see in this example that regular expressions are implemented in Python differently than they are in R. Here are the steps required to perform the task in Python.

Step 1: Import pandas, re, and os library

The three libraries that are used in this script are *pandas*, *re*, and *os*. The *pandas* package is used to read in the data into the script and is used to add a new column to the data frame. The *re* package is used to leverage regular expressions, and the *os* package is used for file system tasks. The code used to do the import is listed as follows:

```
import pandas as pd
import re
import os
```

Step 2: Change working directory

Next, you need to change the working directory to the location on your computer where the *ProductionOrders.csv* file is located using the *chdir()* function from *os*. Here is a code snippet that shows what the code would look like:

```
os.chdir("<path to folder that contains ProductionOrders.csv>")
```

Step 3: Compile the required regular expression

Next, you need to define a regular expression that finds all fuel pumps (561769) and injectors (561394) that were created in AL, LA, or OH and have a model type of A or B. You also want to return only the product number portion of the SKU and not the entire SKU number. You can accomplish this in regular expressions by using *positive lookahead*.

The website www.rexegg.com says that a positive lookahead asserts that what immediately follows the current position in the string is the pattern you are looking for. Let's look at the regular expression in this script to break down what that statement means. Here is the regular expression in question that uses a *positive lookahead*:

```
pattern = "^(561769|561394)(?=(01|22|39)(A|B))
```

The positive lookahead portion of the preceding regular expression is (?=(01|22|39) (A|B)). The structure of a positive lookahead is a regular expression grouped in parenthesis and starts with ?=. The pattern that follows the ?= is the pattern that must follow the preceding pattern in order for the match to pass.

So, the regular expression identified in the *pattern* variable will pass if it

- Starts with 561769 or 561394

- Is preceded by state FIPS codes for AL, LA, or OH

- Has a model type of A or B

This will make more sense when you use it in the function in Step 5. After you define the regular expression, you create a regular expression object named *pattern* by compiling it using the *compile()* function from the *re* package. This is all accomplished using the following code:

```
pattern = re.compile(r"^(561769|561394)(?=(01|22|39)(A|B)\d{1}$)")
```

Step 4: Define a function that returns the monitored products

The name of the custom function that you will create that identifies the monitored products is *monitoredProducts*. Here are the steps to create the function:

1. First, you need to create the shell of the function that accepts the SKU as its only parameter. You do so using the following code:

   ```
   def monitoredProducts(sku):
   ```

2. Next, you attempt to create a match object using the pattern object you defined earlier using the following code:

   ```
   m = pattern.match(sku)
   ```

   If the SKU passed to it matches the regular expression compiled in the *pattern* object, then a match group is returned. If the SKU passed to it does not pass the test, then *None* is returned which is Python equivalent of null.

3. Next, you extract the pattern from the SKU number passed to match function in Step 2 if there is a match. You do so using the following code:

   ```
   if m == True:
       return m.group(0)
   else:
       return "Not Monitored"
   ```

As stated in Step 2, when the code in Step 2 finds a match, a match object with a Boolean value of *True* is returned. Otherwise, the keyword *None* is returned. The preceding *if* statement is used to see if a match object was returned. If one is returned, you get the group in position 0 that represents the complete match. If one is not returned, then the function returns "Not Monitored".

Step 5: Read in the data into Pandas data frame

In this step, you read in the contents of *ProductionOrders.csv* into a data frame named *df* using the following code:

```
df = pd.read_csv("ProductionOrders.csv")
```

Step 6: Create a new column named "Monitored Products"

In this step, you add a column named *Monitored Products* using the following code:

```
df["Monitored Products"] = df["SKU"].apply(monitoredProducts)
```

The preceding code applies the *monitoredProducts()* custom function you created to the SKU column and assigns the results to a new column named *Monitored Products*.

Step 7: Copy the script into Power BI

Here is the complete script:

```
import pandas as pd
import re
import os

os.chdir("<path to ProductionOrders.csv file>")
pattern = re.compile(r"^(561769|561394)(?=(01|22|39)(A|B)\d{1}$)")

def monitoredProducts(sku):

    m = pattern.match(sku)
    if m == True:
        return m.group(0)
```

```
    else:
        return "Not Monitored"

df = pd.read_csv("ProductionOrders.csv")

df["Monitored Products"] = df["sku"].apply(monitoredProducts)
```

Copy the preceding script and paste it into the *Python editor* via *GetData* in Power BI so that you can add it to the Power BI data model.

Summary

In this chapter, we covered several different string manipulation methods that are available to you in Power BI via R and Python. The examples covered may not directly tie to cases you experience in your data wrangling task, but with minor changes to the code, I am sure you can find ways to apply some of these techniques to your situations.

CHAPTER 8

Calculated Columns Using R and Python

A task that is difficult to do in Power Query is adding calculated columns that are based on complex mathematical formulas. That problem does not exist in R or Python. R was built by statisticians, so it was designed to perform complex statistical and mathematical calculations. The same holds true with Python. Python is not only used in data science but also in other computation-intensive fields such as engineering and physics. Like R, Python is optimized to be able to perform calculations that are not possible using M in Power Query. So, in situations where you need to add a calculated column based on a complicated computation in Power BI, you should consider leveraging R or Python. You will not only benefit from the fact that R and Python handle complex computations better but, in many cases, you may find that R and/or Python may have a pre-built function that does the heavy lifting for you.

Let's use a scenario to illustrate this point. Imagine you are a Sr. Analyst for Gee's Trucking, Inc. Gee's Trucking, Inc. recently acquired another trucking company, and they now have over 100 drivers that are operating out of three terminals in metro Indianapolis, IN.

The owner, Gee Troll, wants to calculate the commute distance to work for each of his employees to get an estimate of the average commute distance. Ultimately, he wants to know if any of his drivers would have a shorter commute if they switch terminals. You know that you can easily approximate the distance between two geographical locations using the *Haversine* formula that you learned in college.

The *Haversine* formula can be used to estimate the distance between two points on earth given their longitudes and latitudes. The formula works because the earth is shaped like a sphere.

269

© Ryan Wade 2020

R. Wade, *Advanced Analytics in Power BI with R and Python*, https://doi.org/10.1007/978-1-4842-5829-3_8

You present to your manager four solutions: two solutions using R and two solutions using Python. The two R solutions consist of one custom solution and one solution that leverages a pre-built function. The same holds true for the two solutions in Python. All four solutions require that the addresses are geocoded.

Geocoding is the process of transforming an address to a geographical coordinate made up of a latitude and longitude value. A latitude measures how far north or south a geographical location is from the equator, and the longitude measures how far east or west a geographical location is from the prime meridian.

Here are the basic steps of the process:

1. Generate a Google API key that will be used to communicate with the *Google Geocoding API*.

2. Geocode the addresses using *Google Geocoding API* in R and in Python.

3. Calculate the distance between each employee's home address and the terminal that they work from using a custom function in R and in Python.

4. Calculate the distance between each employee's home address and the terminal they work from using a pre-built function in R and in Python.

Let's start by generating your Google API key.

Create a Google Geocoding API key

Step 1: Log into the Google console

Log into the Google console by going to the following URL: `https://console.cloud.google.com/`. If you don't have a Google account, you will have to create one; otherwise, you should be able to log into the console using your credentials.

Step 2: Set up a billing account

In order to use Google APIs, you need a payment method attached to your account. At the time of writing, Google was offering $300 in credits to test their APIs, which is more than enough to do this demo. If you go over the credits that are allotted to you for free, Google will use the payment method you provided to secure the payment. To set up your payment method, select *Navigation Menu ➤ Billing ➤ Payment method* in the menu as illustrated in Figure 8-1.

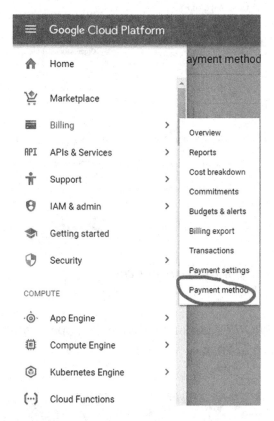

Figure 8-1. *Setting up Google payment method*

Step 3: Add a new project

Next, you need to create a project with geocoding enabled. Make the following menu selections to create a project: *Navigation Menu ➤ IAM & admin ➤ Manage resources*. The menu selection is illustrated in Figure 8-2.

Figure 8-2. *Navigation to Manage resources*

Click the *CREATE PROJECT* button located in the upper right-hand portion of the screen as illustrated in Figure 8-3.

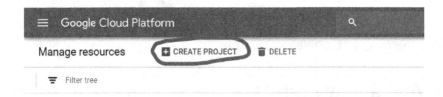

Figure 8-3. *Creating a new Google project*

Populate the form to create the project.

Step 4: Enable Geocoding API

Perform the following steps to enable geocoding in your project:

1. Select *Navigation Menu* ➤ *APIs & Services* ➤ *Library* as illustrated
   in Figure 8-4.

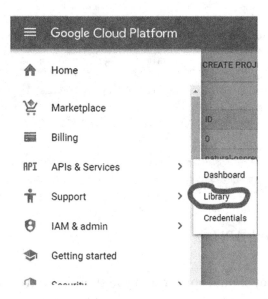

Figure 8-4. *Navigating to the Library*

2. Search for the *Geocoding API* in the *Search for APIs & Services*
   textbox as illustrated in Figure 8-5.

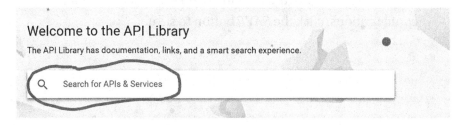

Figure 8-5. *Searching for Geocoding API*

3. Enable the *Geocoding API* by clicking the *Enable* button.

4. Create credentials by doing the following steps:

- Select *Navigation Menu* ➤ *APIs & Services* ➤ *Credentials* in the menu bar.

- Click *Create credentials* and choose *API key* as illustrated in Figure 8-6.

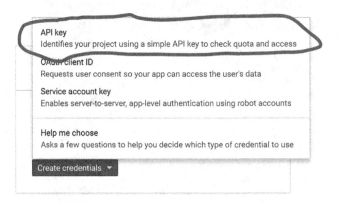

Figure 8-6. *Generating API key*

- An API key should appear, and you should receive the option to restrict the API key. You want to add restrictions to your API key to help prevent unauthorized use. Click *Restrict Key* to set up your restrictions. You will see the web form in Figure 8-7. For this exercise, I recommend to only accept requests from your IP address and restrict the key to the *Geocoding API*. You can do so by using the configurations in Figure 8-7. After you make the configurations, click the *SAVE* button to save.

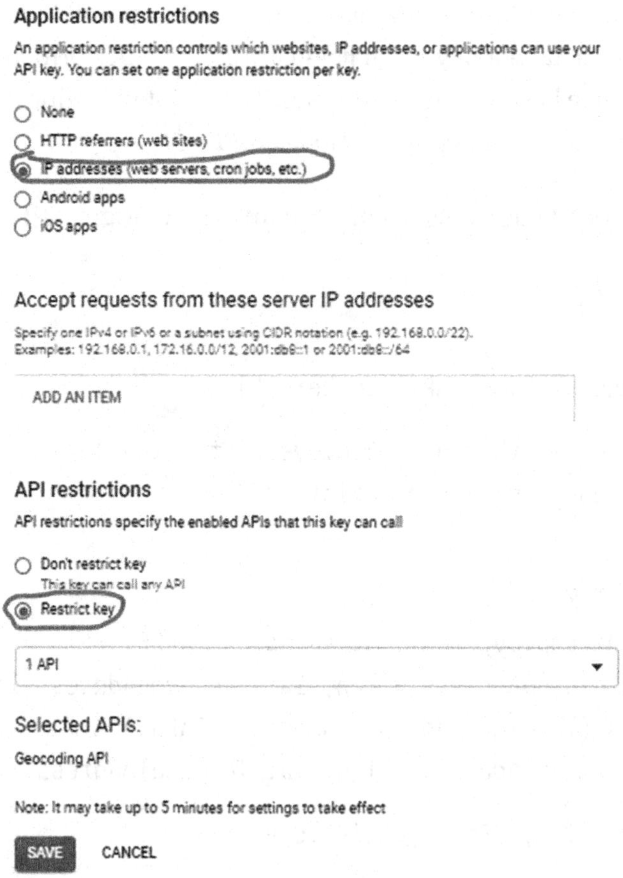

Application restrictions

An application restriction controls which websites, IP addresses, or applications can use your API key. You can set one application restriction per key.

○ None
○ HTTP referrers (web sites)
◉ IP addresses (web servers, cron jobs, etc.)
○ Android apps
○ iOS apps

Accept requests from these server IP addresses

Specify one IPv4 or IPv6 or a subnet using CIDR notation (e.g. 192.168.0.0/22).
Examples: 192.168.0.1, 172.16.0.0/12, 2001:db8::1 or 2001:db8::/64

ADD AN ITEM

API restrictions

API restrictions specify the enabled APIs that this key can call

○ Don't restrict key
   This key can call any API
◉ Restrict key

1 API ▼

Selected APIs:

Geocoding API

Note: It may take up to 5 minutes for settings to take effect

SAVE   CANCEL

Figure 8-7. *Configuring API restrictions*

5. You now have an API key configured with restrictions. Save your key in a safe place. If you misplace it, you can go back to the Google console to get a copy of your API key.

Geocode the addresses using R

Now that you have your Google API key configured, you can use R to leverage the API to geocode the addresses in the *EmployeeList.csv* file. The process of geocoding will convert each address to a coordinate that represents a point on the earth, which is defined by a *longitude* and *latitude* value. The *longitude* value tells you how far east or west you are from the *prime meridian,* and the *latitude* value tells you how far north or south you are from the *equator*. Having addresses in this format makes it possible to perform *spatial analytics*. The type of spatial analytics that you are going to perform in this exercise is estimating the distance between two addresses. The script in Listing 8-1 shows how to geocode addresses in R by leveraging the *Google API*.

Listing 8-1. Geocoding addresses with R using the Google API

```
library(tidyverse)
library(ggmap)

register_google(key = "<put API key here>")

setwd("<path to folder that host EmployeeList file>")
EmployeeList <- read_csv("EmployeeList.csv")

EmployeeList <-
    EmployeeList %>%
    mutate_geocode(EmployeeAddress, sensor = FALSE) %>%
    rename(lon_EmployeeAddress = lon, lat_EmployeeAddress=lat) %>%
    mutate_geocode(TerminalAddress, sensor = FALSE) %>%
    rename(lon_TerminalAddress = lon, lat_TerminalAddress = lat)

write_csv(EmployeeList, "EmployeeList.csv")
```

At this point of the book, you have a decent understanding of R, so I won't be as pedantic with my explanations as I was early on in the book. Here are succinct explanations of the script broken out in logical steps:

1. Start by importing the required package. A new package, *ggmap*, is introduced in this script. The main use for the *ggmap* package is visualizing spatial data, but you will be using it to geocode the addresses.

2. You use the Google Geocoding API service to geocode the addresses. In order to do so, you need to register your API key using the *register_google()* function.

3. Next, you read in the *Employee.csv* file.

4. Next, you use *dplyr* and the *mutate_geocode()* verb from *ggmap* to add the geocode information to your data set. Like the *mutate()* verb from *dplyr*, *mutate_geocode()* is used to add, or project, new columns to your data frame. The *mutate_geocode()* specifically adds the coordinates of the addresses passed to it. It adds a column named *lon* that contains the longitudes of the addresses passed to it and a column named *lat* that contains the latitudes of the addresses passed to it.

5. Next, you use the *rename()* verb from *dplyr* to give the *lon* and *lat* columns a more meaningful name.

You have two options that you can use to rename columns using *dplyr*, *rename()*, and *select()*. What makes *rename()* different than *select()* is that rename keeps all other columns in your data set and not only the ones you are renaming, whereas *select()* only keeps the columns you have in your select verb whether it be columns you are renaming or columns you are selecting from your data frame.

---

6. You do Steps 4 and 5 for the *TerminalAddress*.

In this exercise, you were able to communicate with the *Google API* via R to geocode the addresses. Making API calls with Power Query is not as easy when you are not on premium compacity. If you are using *Power BI Premium,* you have the option to use *Microsoft Cognitive Services,* but that is it. With R and Python, you don't have that limitation. In this example, we used the *Google API* to geocode the addresses, but later in the book, we will show how you can leverage other APIs such as *IBM Watson Natural Language Understanding.* In the next section, you will learn how to geocode your addresses in Python.

Geocode the addresses using Python

In the previous section, you geocoded your addresses using the *Google API* via R. In this section, you will do the same using Python. As stated in the previous section, working with APIs in Power Query is much harder unless you are using the relatively expensive *Power BI Premium.* The following script in Listing 8-2 shows how to geocode addresses in Python by leveraging the *Google API*.

Listing 8-2. Geocoding addresses with Python using the Google API

```python
import pandas as pd
import os
from geopy import geocoders

g_api_key = "<API Key>"
g = geocoders.GoogleV3(g_api_key)

os.chdir("<path to folder where the EmployeeList file is located>")
EmployeeList = pd.read_csv("EmployeeList.csv")

EmployeeList["EmployeeAddressGC"] = \
    EmployeeList["EmployeeAddress"].apply(g.geocode)

EmployeeList["lat_EmployeeAddress"] = \
    EmployeeList["EmployeeAddressGC"].apply(lambda x: x.latitude)

EmployeeList["lon_EmployeeAddress"] = \
    EmployeeList["EmployeeAddressGC"].apply(lambda x: x.longitude)

EmployeeList["TerminalAddressGC"] = \
    EmployeeList["TerminalAddress"].apply(g.geocode)

EmployeeList["lat_TerminalAddressGC"] = \
    EmployeeList["TerminalAddressGC"].apply(lambda x: x.latitude)

EmployeeList["lon_TerminalAddress"] = \
    EmployeeList["TerminalAddressGC"].apply(lambda x: x.longitude)

cols = ["EmployeeAddressGC", "TerminalAddressGC"]
EmployeeList.drop(cols, inplace=True)

EmployeeList.to_csv("EmployeeList.csv", index = False)
```

Because you are at the point in the book where you should have developed a good understanding of Python, the code explanations will not be as verbose as they were earlier in the book. Here is the explanation of the preceding Python script broken out in logical steps:

1. You start with your library imports. The new module that is being used in this script is the *geocoders* module from the *geopy* library. This module is used to take care of the geocoding.

2. Next, you create an object, g, that is based on the *Google API* class, *GoogleV3*. This object will be used to query the *GoogleV3* API class to geocode the addresses.

3. Next, you read in the *Employee.csv* file.

4. The data set contains two addresses that need geocoding, the *EmployeeAddress* address and the *TerminalAddress* address. First, the *EmployeeAddress* is geocoded by using the apply method of the *EmployeeAddress* column to apply the geocode function from object *g* using the following line of code:

```
EmployeeList["EmployeeAddressGC"] = EmployeeList
["EmployeeAddress"].apply(g.geocode)
```

The preceding line of code creates a *location.Location* object that contains the geocoding information you need plus more. The object is stored in the *EmployeeList["EmployeeAddressGC"]* field. You will extract information from that object in subsequent steps.

5. Next, you create the *lat_EmployeeAddress* column by extracting information from the *EmployeeAddressGC* column created in Step 4. As mentioned earlier, if Google was successful at geocoding the address that you sent to it, Google will send you a *location.Location* object that contains the address, latitude, and longitude. You are interested in the latitude and longitude. You can extract that information from the *location.Location* by applying a *lambda* function to the *EmployeeAddressGC* field.

A *lambda* function is a nameless function that only has one expression. They are identified with the keyword *lambda,* followed by the arguments, followed by the expression as illustrated here: `lambda arguments: <expression>`. They provide a succinct way of applying functions to *pandas* data frames.

Use the following lambda function to extract the latitude from *EmployeeAddressGC*: `lambda x: x.latitude`. The x in the lambda function represents the *EmployeeAddressGC* field. The code `x.latitude` enables you to access the latitude property in x.

6. Next, you create the *lon_EmployeeAddress* column by using the same steps that you performed in Step 5 but replacing latitude with longitude.

7. Next, you repeat Steps 4, 5, and 6 for the *TerminalAddress* field.

8. Next, you remove the unneeded columns from your data frame. You only needed *EmployeeAddressGC* and *TerminalAddressGC* to extract the latitude and longitude of the employee address and terminal address. Those fields are no longer needed, so you can remove them. An easy way to do that is using the *drop* method of the *EmployeeList* data frame. That is done using the following code:

```
cols = ["EmployeeAddressGC", "TerminalAddressGC"]

EmployeeList.drop(cols, inplace=True)
```

The *cols* variable contains the list of columns you want to drop. By default, the drop method will not act directly on the data frame. If you want it to act directly on the data frame and permanently remove the specified columns, you need to set the *inplace* argument to *True.*

In this exercise, you were able to communicate with the *Google API* via Python to geocode the addresses. You were able to perform a task relatively easy in Python that would have been much harder in Power Query. In the next section, you will learn how to use the geocoded information to estimate distance in miles using a custom function built in R.

Calculate the distance with a custom function using R

In the previous exercise, you setup your Geocoding API and you geocoded your addresses. Now the data is in the proper format to calculate the distance. Listing 8-3 is the R script that calculates the distances in miles between the employee addresses and their terminal.

Listing 8-3. Calculating distance in R using a custom function

```
library(tidyverse)

ComputeDist <-
  function(pickup_long, pickup_lat, dropoff_long, dropoff_lat) {
    R <- 6371 / 1.609344
    delta_lat <- dropoff_lat - pickup_lat
    delta_long <- dropoff_long - pickup_long
    degrees_to_radians = pi / 180.0
    a1 <- sin(delta_lat / 2 * degrees_to_radians)
    a2 <- as.numeric(a1) ^ 2
    a3 <- cos(pickup_lat * degrees_to_radians)
    a4 <- cos(dropoff_lat * degrees_to_radians)
    a5 <- sin(delta_long / 2 * degrees_to_radians)
    a6 <- as.numeric(a5) ^ 2
    a <- a2 + a3 * a4 * a6
    c <- 2 * atan2(sqrt(a), sqrt(1 - a))
    d <- R * c
    return(d)
  }

setwd("<path to folder where the EmployeeList file is located>")
EmployeeList <- read_csv("EmployeeList.csv")

EmployeeList <-
  EmployeeList %>%
  mutate(
    `Custom Distance Function Results` =
      round(
```

```
ComputeDist(
    lon_EmployeeAddress,
    lat_EmployeeAddress,
    lon_TerminalAddress,
    lat_TerminalAddress
),1
    )
)
```

The code is unpacked into the following logical steps:

1. First, you need to load the required package. The only one needed in this script is the *tidyverse* metapackage because most of the functionality needed is already included in base R.

2. Next, you create a custom function named *ComputeDist* to calculate the distance. The *ComputeDist* custom function requires four parameters: the latitude and longitude of the Employee and Terminal addresses, respectively.

   Notice how involved the calculation is. Performing this type of math in Power Query using M would be much harder to do and would not be as performant as it is in R. I will not explain how this calculation works because it is beyond the scope of this book. The purpose of this example is to show how R does a much better job than M at performing math-intensive calculations.

3. Next, you load the *EmployeeList.csv* file.

4. Next, you use the *mutate()* verb from *dplyr* to add a new column named *Custom Distance Function Results* to the *EmployeeList* data frame. The name *Custom Distance Function Results* was used to make it clear that a custom function was used to calculate the distance. The four required arguments are passed to the function, and the *round()* function is used to round the returned value to one decimal point.

5. The last thing you do is copy the entire script and add it to the Power BI data model via the R script editor in *GetData*.

In this exercise, you were able to create a calculated column based on a custom function in five easy steps. The nice thing about using a custom function is that you were able to abstract the complexities of the math needed to perform the calculation in a custom function and you were able to invoke it whenever it is needed. The result is code that is more readable and easier to maintain.

Calculate the distance with a custom function using Python

In a previous step, you geocoded your addresses using the Google Geocoding API. Now the data is in the proper format to calculate the distance. Listing 8-4 is the Python script that you will use to estimate the distance in miles between the employee address and their terminal.

Listing 8-4. Calculating distance in Python using a custom function

```python
import pandas as pd
import numpy as np
import os
from math import cos, sin, atan2, pi, sqrt, pow

def ComputeDist(row):
    R = 6371 / 1.609344 #radius in mile

    delta_lat = row["lat_TerminalAddressGC"] -
    row["lat_EmployeeAddress"]

    delta_lon = row["lon_TerminalAddress"] -
    row["lon_EmployeeAddress"]

    degrees_to_radians = pi / 180.0
    a1 = sin(delta_lat / 2 * degrees_to_radians)
    a2 = pow(a1,2)
    a3 = cos(row["lat_EmployeeAddress"] * degrees_to_radians)
    a4 = cos(row["lat_TerminalAddress"] * degrees_to_radians)
    a5 = sin(delta_lon / 2 * degrees_to_radians)
    a6 = pow(a5,2)
```

```
a = a2 + a3 * a4 * a6
c = 2 * atan2(sqrt(a), sqrt(1 - a))
d = R * c
return d
```

```
os.chdir("<path to folder where the EmployeeList file is>")
EmployeeList = pd.read_csv("EmployeeList_Python.csv")
```

```
EmployeeList["Custom Function"] = \
    EmployeeList.apply(lambda row: ComputeDist(row), axis=1)
```

Let's unpack the preceding code into logical steps:

1. First, you need to load the required library. What's new in this script are some functions from the math library. These functions are used to calculate the distance.

2. Next, you create custom function named *ComputeDist()* that estimates the distance in miles between two geographical points. The function requires one parameter, and that is the entire row from the data frame. You will extract the information from the row that is needed to perform the calculation. The calculation is very involved, and performing this type of math in Power Query using M would be harder to do and not as performant as it is in Python. I will not explain how this calculation works because it is beyond the scope of this book. The purpose is to show how Python does a much better job at performing math-intensive calculations than M.

3. Next, you load the *EmployeeList.csv* file.

4. Next, you use the *apply* method to apply a *lambda* function on the *EmployeeList* data frame. The complete line of code is as follows:

```
EmployeeList["Custom Function"] = (
    EmployeeList.apply(lambda row: ComputeDist(row), axis=1))
```

The use of the parentheses in the preceding code enables you to continue code on a subsequent line, making the code more readable. The subsequent lines must follow the indentation rules defined in *PEP8*. *PEP8* outlines the rule for formatting your Python code. You can read more about *PEP8* at this URL: `https://realpython.com/python-pep8/`.

You were exposed to *lambda* functions earlier in the chapter. The *lambda* functions you used were applied to a specific column in a data frame, but the *lambda* function in this example is being applied to the whole data frame. When you apply a *lambda* function to a data frame, you have the option of telling pandas if you want to apply it to the *0 axis* or *1 axis*. If you apply it to the *0 axis*, then the lambda function will be applied top-down, but if you apply it to the *1 axis*, it will be applied left to right. Let's make things simpler with an example.

Let's create a small data frame with the code beneath to illustrate how to apply a *lambda* function to a data frame:

```
df = pd.DataFrame((
    {'A':[10, 20, 30], 'B':[15, 5, 10]}))
```

The preceding code produces the following data frame:

	A	B
0	10	15
1	20	5
2	30	10

If you want to get the sum for each column, you need to work along the *0 axis* so you can operate from top to down. You do so using the following code:

```
df.apply(sum, axis = 0)
```

The code produces the following output:

A	60
B	30

By setting the axis to 0, the code will work "top-down" along each column, and it will sum up all the elements in each respective column. If you set the axis to 1, it will work left to right and for each row, and it will sum up all the elements in each of the rows. So, if you change the axis to 1 as illustrated in the following code

```
df.apply(sum, axis = 1)
```

It will produce the following output:	
0	25
1	25
2	40

Now you understand how lambda functions work when applied to a data frame. Let's revisit the code in this step from the original script. Here's the code:

```
EmployeeList["Custom Function"] = (
    EmployeeList.apply(lambda row: ComputeDist(row),
    axis=1))
```

The preceding code is applying the lambda function to the 1 axis. In other words, it is working left to right on each row. The entire row is passed to the *ComputeDist()* function. The *ComputeDist()* function extracts the information it needs for the calculation from the row object that was passed to it. Note that you are not limited to aggregation functions when you use the apply method against a data frame.

5. Copy the preceding script in Listing 8-4 to add the data frame that is generated by the script to the Power BI data model via the *Python script editor* in *GetData*.

Like in the previous R exercise, you were able to create a calculated column based on a custom function in five easy steps. Like in R, you were able to gain the benefit of having code that is more readable and easier to maintain when you use custom functions to create calculated columns in Power BI using Python.

Calculate distance with a pre-built function in Power BI using R

In the previous two sections, you learned how to develop custom functions in both R and Python to efficiently perform computations that are math-intensive. Even though it was not that hard to implement a math-intensive custom function, there are often much easier solutions. At the time of this writing, the R programming language has over 15K packages, and the Python programming language has well over 110K libraries. Often you can find a solution to your problem in one of the many packages and libraries that are available in R and Python, respectively.

That is the case with the calculating distance between two geographical points. Instead of refactoring a mathematical formula in R code or Python code, you can leverage a pre-built function. In this exercise, you will learn how to leverage the *geosphere* package in R to perform the distance calculation. The complete script you will use is in Listing 8-5.

Listing 8-5. Calculating distance in R using a pre-built function

```
library(tidyverse)
library(geosphere)

setwd("<path to folder where the EmployeeList file is located>")
EmployeeList <- read_csv("EmployeeList.csv")

EmployeeList <-
    EmployeeList %>%
    rowwise() %>%
    mutate(
```

```
`Geosphere Function Result` =
    distHaversine(
        c(lon_EmployeeAddress, lat_EmployeeAddress),
        c(lon_TerminalAddress, lat_TerminalAddress)
    ) / 1609.34
)
```

Let's unpack the code into logical steps:

1. First, load the required packages. The new package in this script is the *geosphere* package. You will leverage this package to perform the spatial calculations.

2. Next, you load the contents of the *EmployeeList.csv* file into the *EmployeeList* data frame.

3. Next, you add a new column named *Geosphere Function* to the *EmployeeList* data frame. You are using the chaining method in *dplyr* to create a workflow that adds a column to the data frame that calculates the distance based on the *distHaversine* function. The first step of the workflow tells *dplyr* that you are starting with the original data frame. The second step of the workflow passes the data frame to the *rowwise()* function. The *rowwise()* function is necessary because the *distHaversine()* function that is used in the subsequent step is not vectorized.

A *vectorized* function works on a whole vector at the same time. Columns in a data frame are vectors. When you apply a *vectorized* function against a vector, you don't need to implement in looping logic.

The *rowwise()* function tells all subsequent tasks in the workflow to only consider the elements in the current row in the calculation. The last step performs the calculation using the *distHaversine()* function from the geosphere package. The *distHaversine()* function returns the distance in meters, so multiplying the results by 0.000621371 converts the value to miles.

4. As always, you need to copy the entire script in Listing 8-5 to the Python editor in Power BI so that you can add the resulting data set to the Power BI data model.

In this exercise, you were able to leverage a pre-built function from the *geosphere* package that abstracted the complexity of the calculation from you. The function that was used was the *distHaversine()* function. All you had to do was pass the function the required parameters and it performed the calculation for you. In the next exercise, you will use similar functionality in Python.

Calculate distance with a pre-built function in Power BI using Python

The following Python script (Listing 8-6) uses a pre-built function to calculate the distance.

Listing 8-6. Calculating distance in Python using a pre-built function

```python
import pandas as pd
import os
from haversine import haversine

os.chdir(r"<file path to folder that contains EmployeeList.csv>")
EmployeeList = pd.read_csv("EmployeeList.csv")

def useHaversine(row):
    point_one = \
        (row["lat_EmployeeAddress"], row["lon_EmployeeAddress"])

    point_two = \
        (row["lat_TerminalAddress"], row["lon_TerminalAddress"])

    return haversine(point_one, point_two, unit="mi")

EmployeeList["Haversine Function Result"] = \
    EmployeeList.apply(lambda row: useHaversine(row), axis=1)
```

Let's unpack the preceding code into logical steps:

1. First, you need to load the required libraries. The new library in this script is the *haversine* library. This library is leveraged to perform the spatial calculations.

2. Next, you load the contents of the *EmployeeList.csv* file into the *EmployeeList* data frame.

3. Next, you define the *useHaversine()* function. The function accepts the entire row as its input. Information is easily extracted from the row passed to the function using methods you have already learned. The *useHaversine()* custom function in this example uses the *haversine()* function from the *haversine* library to calculate the distance. You pass three parameters to the *haversine()* function: the coordinates of the *location_one*, the coordinates of *location_two*, and the units that you want the distance in. The *location_one* and *location_two* variables need to be in a form of a *tuple* with the first element in the tuple being the *latitude* and the second element in the tuple being the *longitude*.

   The reason why the *Haversine()* function from the *Haversine* library is wrapped in the *useHaversine()* custom function is because the *Haversine()* function is not vectorized. If the *Haversine()* function was vectorized, it would be able to operate on all the rows in the data frame in parallel, but since it is not, it can only operate on one row at a time. That is why a single row is passed to the *useHaversine()* function.

4. Next, create a column named *Haversine Function Result* that will contain the distance between the Employee's address and the Terminal using the following code:

```
EmployeeList["Haversine Function Result"] = Employee
List.apply(lambda row: useHaversine(row), axis=1)
```

   A *lambda* function is applied to the *EmployeeList* data frame along *axis 1*. As you learned earlier in the custom function example, this enables you to pass the entire row to the *useHaversine()* function.

5. Last, but not least, copy the script and paste it into the Python editor in Power BI so that the resulting data frame can be added to the Power BI data model.

In this exercise, you were able to leverage a pre-built function from the *Haversine* library that abstracted the complexity of the calculation using the *Haversine* formula from you. The function that was used was the *Haversine()* function. Just like in R, you were able to leverage a pre-built function that performs the mathematical calculations for you using Python. All you had to do was make a function call.

Summary

This chapter illustrated two very important techniques. The first technique was how to use R and Python to create calculated columns based on a complex mathematical formula. The example in this chapter showed how R and Python handle complex math much better than M does in Power Query. The same would hold true when comparing R and Python to DAX for most scenarios. The second technique showed how to use pre-built functions from one of the many packages and libraries in R and Python, respectively. The exercise used pre-built R and Python functions based on the *Haversine* formula, but chances are that one of the 15,000+ R packages and one of the 110,000+ Python libraries have a pre-built function that you can use to abstract the complexity of the distance calculation from you for the vast majority of your calculation needs.

The ability to use R and Python in Power BI to create calculated columns can be used to overcome many other Power Query limitations. In addition to performing complex calculations, you can also use R or Python to

- Create a calculated column that imputes values for missing data points based on algorithms that are more accurate than common methods such as using the mean or the mode

- Create a calculated column based on complex string manipulations as you learned in *Chapter 7*

- Create a calculated column that is based on a machine learning algorithm that you will learn in Chapter 9

- And many more

The possibilities are numerous!

PART IV

Machine Learning and AI in Power BI Using R and Python

CHAPTER 9

Applying Machine Learning and AI to Your Power BI Data Models

Having a mature enterprise business intelligence solution has become the norm for many organizations. A considerable number of firms have well-governed Enterprise Data Warehouses (EDWs) that are updated via sophisticated ETL processes. They use reporting tools such as SQL Server Report Services (SSRS) and Power BI to help the business gain valuable insights about their business. Organizations that fall in this category have done a great job of leveraging BI, and they are primed to introduce *artificial intelligence (AI)* to their data strategy. Here are a few ways AI can be used to enhance a mature business intelligence system:

- A bike shop can use a *logistic regression* model to add an attribute to their customer dimension table to identify customers that are likely to buy a bike within the next year.

- A retail company can use the *k-means clustering* algorithm to add an attribute to their customer dimension table that segments their customers for marketing purposes.

- An online retailer can use a *logistic regression* model to add an attribute to their customer dimension table that can be used to identify customers that are likely to churn.

- A restaurant can use *sentiment analysis* to add a sentiment attribute to their customer dimension table based on the customer's reviews.

© Ryan Wade 2020

R. Wade, *Advanced Analytics in Power BI with R and Python*, https://doi.org/10.1007/978-1-4842-5829-3_9

These types of AI-generated attributes can be very valuable and are not easily calculated using traditional means. Once these AI-generated attributes are added to your data model, they can be used like any other attribute in your dimension tables. They can be used in your Power BI visuals, tables, and matrixes.

In this chapter, you will learn how to leverage AI in your Power BI data models. This chapter focuses on self-service methods, but later in the book, you will learn methods to operationalize your AI models at scale without the need of Power BI Premium.

Power BI Premium has a feature called *AI Insights* that facilitates applying AI to your Power BI data models. This book will not cover *AI Insights* because of the price associated with it. All examples in this book can be done without the need for premium compacity.

The chapter will introduce you to multiple ways you can apply AI to your Power BI data. You will learn

- How to score your Power BI data using *custom machine learning* models built in R and Python

- How to use *Microsoft Cognitive Services* to apply sentiment analysis to your Power BI data model

- How to use *IBM Watson Natural Language Understanding* to apply tone analysis to your Power BI data model

This chapter focuses on self-service methods, but later in the book, you will learn how to operationalize some of the techniques outlined in this chapter for enterprise solutions. Let's get started with scoring your data using custom models built in R and Python!

Apply machine learning to a data set before bringing it into the Power BI data model

Here is the scenario. You work for a real estate analytics firm in the city of Boston, MA. Your data scientist was asked to help you predict median house prices in Boston. She helped you by developing a machine learning model in both R and Python using the

Boston housing data set from Kaggle. She saved the models for you in a shared directory. You will use those models to score a new data set that contains a list of homes whose prices you want to predict.

The new data set contains 14 variables, but the model your data scientist developed only uses 4 of them. Here are the four variables that the model uses:

- CRIM: Per capita crime rate by town

- RM: Average number of rooms per dwelling

- TAX: Full-value property-tax rate per $10,000

- LSTAT: % lower status of the population

You will create both an R and Python script for Power BI that will perform the scoring. First, you will start with R.

Predicting home values using R

Applying machine learning models developed by a data scientist to your Power BI data models is straightforward. The steps to do so in R are outlined as follows. This book will not cover how to build the actual models because that is outside the scope of this book. The focus will be on how to apply a model that was created for you. Here are the steps.

Step 1: Have the data scientist share the model with you

An R model is an object that can be saved to disk. After the data scientist builds the model, she can save the model to disk and share it with you in the same manner that she would share any other file. The process used to serialize the model to disk is very easy to do in R and can be accomplished with one line of code. Here's the code:

```
saveRDS(model, "model.rds")
```

The preceding code will take the model that the data scientist created that is stored in the variable named *model* and save it to disk as a **.rds* file in your working directory using the *saveRDS()* function. The *saveRDS()* function is part of the base package.

Step 2: Load the tidyverse package

You will start developing the script to score the data in this step. You only need the *tidyverse* package in this script, but most of what you will do can be handled with R's base functionality. The *tidyverse* metapackage is loaded using the following code:

```
library(tidyverse)
```

Step 3: Load the model object and the data set to be scored

You do so using the following code:

```
model <- readRDS("./Model/model.rds")
boston_housing <- read_csv("./Data/BostonHousingData.csv")
```

The *readRDS()* function allows you to deserialize the model from disk so that you can use it in your script. You used the *read_csv()* function from the *readr* package to read in the data that needs to be scored into an R data frame named *boston_housing*.

Step 4: Subset the data frame so that it only contains the columns needed for the model

The data set has 13 variables, but you only need 4 of them for the model. You subset the variables you need from the *boston_housing* data frame using the following code:

```
model_data <- boston_housing[,c("crim","rm","tax","lstat")]
```

The empty space after the bracket followed by the comma tells R that you want to include all rows. The character vector that follows the comma tells R the columns that you want to subset. So, the preceding code produces a data frame that contains all the rows from the *boston_housing* data frame but only the columns listed in the character vector. Now you have a data set that only contains the data needed for the model via the *model_data* data frame.

Step 5: Apply the model to your data set to predict the median home values

You accomplish this task with the following code:

```
pred_medv <- predict(model, model_data)
```

The preceding code returns a vector that contains the median home value predictions for each observation that was passed to it.

The formal definition of an observation is a recording of qualities and quantities of an observable phenomenon in the natural world. So, in this example, the observable phenomenon is the information you have about each house you want to predict.

Note that the data frame passed to the *predict()* function must contain the variables (columns) that are needed in the model.

Step 6: Add the predictions to the original data set

The code from Step 5 only returned a vector of data, which is meaningless by itself. You need to combine it back to the original data set. You can easily do so using the following code:

```
final_output <- cbind(model_data,pred_medv)
```

The code uses the *cbind()* function to attach a column to the end of the *model_data* data frame, then assigns the resulting data frame to a variable named *final_output*. The *c* in *cbind()* stands for column. In this case, it binds the vector *pred_medv* to the *model_data* data frame as a new column and gives the new column the name of the vector, which in this case is *pred_medv*.

Step 7: Copy the entire R script into the Power BI data model

The final script is below in Listing 9-1.

Listing 9-1. R Script to predict Boston housing prices

```
library(tidyverse)

model <- readRDS("./Models/model.rds")
boston_housing <- read_csv("./Data/BostonHousingInfo.csv")

model_data <- boston_housing[, c("crim","rm","tax","lstat")]
pred_medv <- predict(model, model_data)

final_output <- cbind(model_data, pred_medv)
```

Like before, you copy the script, then go to *GetData* ➤ *More* ➤ *Other* in Power BI. Next, you click *R script* and paste the script inside the *R editor* in Power BI. After that, you click *OK*. The *final_output* data frame will be exposed to you by Power BI as a data set that you can bring into the Power BI data model. You can combine it with other data sets in the Power BI data model, create measures that use the predictions, or create visualizations using the scored data.

Predicting home values using Python

In this section, you will also predict Boston house prices, but you will do so using Python instead of R. As in the R example, the focus will be on how to apply a model that was created for you because the creation of the model is outside the scope of the book. Here are the steps in Python.

Step 1: Have the data scientist share the model with you

Like with R, a Python model is an object that can be saved to disk so it can be shared with you in the same manner that the R model was shared earlier. The process to serialize the model to disk in Python is very easy to do, and it can be accomplished with one line of code after importing the *joblib* library. Here's the code:

```
import joblib
joblib.dump(model,"model")
```

The preceding code takes the model that the data scientist created for you and saves it to disk in the current working directory using the *joblib.dump()* function from the *joblib* library. After the model has been saved to disk, the data scientist can share the model with you in a shared repository such as *OneDrive* or *Dropbox*. A process known as *MLOps* may be used in a more sophisticated shops so the model may be shared with you via a repository such as *GitHub* or *Azure DevOps* in that scenario.

Step 2: Load the necessary Python libraries needed for the script

This is the step where you actually start building the script that will do the scoring in Power BI. Here is the code needed for your import section of your Python script:

```
import os
import joblib
from sklearn import preprocessing
import pandas as pd
```

The *joblib* library is used to deserialize the model that the data scientist shared with you, the *preprocessing* module from *scikit-learn* provides you with tools to transform your data to get it ready for scoring, and the *pandas* library will handle your data frame needs.

Step 3: Load the model object and the Boston homes data set

You can do so using the following code:

```
model = joblib.load("../Models/model_Python.pkl")
boston_housing = pd.read_csv("../Data/BostonHousingInfo.csv")
```

The *joblib.load* function is very similar to the *readRDS* function we used in R. It allows you to deserialize the model from disk so that you can use it in your script. You used the *read_csv* method in *pandas* to read in the data that needs to be scored into a *pandas* data frame. Note that both file paths used are prefix with a *".."*. Prefixing your file path with a *".."* enables you to start your relative file path referencing one level up from your currently working directory. The *Models* and *Data* folder are at the same level as the *Python* folder which is the folder the current working directory is set to. Using the *".."* allows you to go one level up to the *Chapter09* folder which enables you to access the contents in *Models* folder and *Data* folder.

Step 4: Extract the information needed from the bost_housing data frame

The data set that you are using has 13 variables, but you only need 4 of them for the model. After you extract the variables from the *bost_housing* data frame that you need, you will need to standardize them.

Standardization is often required by models in *scikit-learn*. It is usually recommended in machine learning when working with variables that have different scales, as do the variables used in this example. When you are developing a model using variables of different scales, variables with the bigger values will carry more weight and can cause bias in your model. The *scale()* function from the *preprocessing* module puts all the variables on the same scale, which alleviates that problem. Here is the URL to a YouTube video that further explains standardization: `www.youtube.com/watch?v=sh_tLn1phfc`.

You subset and standardize the data you need for the model using the following code:

```
model_data = preprocessing.scale(
    boston_housing[["crim","rm","tax","lstat"]])
```

Let's first break down the code passed to the *scale()* function, then explain what the *scale()* function does. Here's the code passed to the *scale()* function:

```
boston_housing[["crim","rm","tax","lstat"]].
```

This code subsets the *boston_housing* data frame so that it only includes the columns defined in the list contained inside the brackets that are used to subset the data frame. Next, the resulting data frame is passed to the *scale()* function where it will be standardized. Once standardized, all the variables will be on the same scale.

Step 5: Apply the model to the preprocessed data to predict the median home values

You accomplish this task with the following code:

```
pred_medv = model.predict(model_data)
```

The preceding code returns an array that contains the median home value predictions for each observation that was passed to it. The resulting output is assigned to a variable named *pred_medv*. Note it is important that the *model_data* data frame contains the variables that are needed in the *model* model.

Step 6: Add the predictions to the original data set

The code from Step 6 only returned an array of data which is meaningless by itself. You need to combine it back to the original data set. You can easily do so using the following code:

```
final_output = boston_housing.loc[:,["crim","rm","tax","lstat"]]
final_output["pred_medv"] = pred_medv
```

You want to name the data set that will be passed to Power BI *final_output*. To do so, you start by creating a data frame named *final_output* that is a subset of the *boston_housing* data frame. You only need the *crim, rm, tax*, and *lsat* columns from the *boston_housing* data frame.

You will use the iloc method of the *boston_housing* data frame to subset those columns. The *iloc* method allows you to subset both the rows and the columns of a data frame at the same time. The first argument subsets the rows, and the second argument subsets the column. In this example, you want all rows from the *boston_housing* data frame so you tell *pandas* that with the : symbol. Note if you wanted the first three rows, you would use 0:2. You put the integer position of the row you want to start with on the left of the semicolon and the integer position of the row you want to end with on the right of the semicolon. You would start your range with 0 because python uses a 0-based index. The next argument is used to tell *pandas* what columns you want to return. You can use : if you want to return all columns, but if you want to return a subset of the column, you can use a *list* that contains the columns you want to return as we did in this example. Lastly, you append a new column named *pred_medv* to the *final_output* data frame using the following code: `final_output["pred_medv"] = pred_medv`.

Step 7: Copy the entire Python script into the Python script editor in Power BI

The final script is shown in Listing 9-2.

Listing 9-2. Python script to predict Boston housing prices

```
import os
import pandas as pd
import joblib
from sklearn import preprocessing

model = joblib.load("../Models/model_Python.pkl")
boston_housing = pd.read_csv("../Data/BostonHousingInfo.csv")

model_data = preprocessing.scale(boston_housing[["crim","rm","tax","lstat"]])

pred_medv = model.predict(model_data)

final_output = boston_housing.loc[:,["crim","rm","tax","lstat"]]
final_output["pred_medv"] = pred_medv
```

Like in the examples before, you go to *GetData* ➤ *More* ➤ *Other,* then click *Python script.* Next, you paste the preceding script in the *Python editor,* then click *OK.* The *final_output* data frame will be exposed to Power BI as a data set that can be brought into the Power BI data model.

If needed, the field with the prediction can be further enhanced in Power Query to make it more valuable before it is added to the data model. For instance, if a real estate company was using this model, they could put the predicted house prices into bins. For instance, the houses with predicted prices less than 100K could go in the *<100K* bin, the houses with predicted prices between 100K and 200K can be put into the *100–200K* bin, houses with predicted prices between 200K and 300K can be put into the *200–300K* bin, and so on. Doing this type of binning may help the business gain valuable insights about their situation.

Using pre-built AI models to enhance Power BI data models

Developing *AI models* is not a trivial task. The amount of time it takes to develop a model that generalizes well against real data can be resource- and time-intensive. It often requires a team of data scientists and data engineers to accomplish such a task. Fortunately, big software vendors such as *Microsoft* and *IBM* have developed sophisticated *AI models* that are easily accessible via *API calls*. These models are based on years of research that would be hard for you to replicate on your own.

In this section, you will learn how these models can be used to enhance your Power BI data models. You will learn how to use *Microsoft Cognitive Services* to perform *sentiment analysis*.

Sentiment analysis is used to detect how positive or negative the sentiment of a piece of text (sentence, paragraph, tweet, etc.) is. In *Microsoft Cognitive Services*, this is done by assigning a numerical score between 0 and 1. The closer the score is to 0, the more negative the text is, and the closer the score is to 1, the more positive the text is.

You can use *sentiment analysis* to enhance your Power BI data models by using it to add attributes to your dimension tables. For instance, if you are a restaurant, the average sentiment scores of your customer reviews can be the basis of an attribute in your customer dimension table. Such an attribute can be used to find associations between things such as *sentiment* and customer attrition. In a restaurant example, you may discover that the majority of low sentiment scores were associated with customers who purchased particular items on the menu.

Let's go over the steps needed to add *sentiment analysis* to your Power BI model using *Microsoft Cognitive Services*. The example will only be done in using Python because R is not currently supported. You can also perform this task using *AI Insights* via *Power BI Dataflows,* but that requires *Power BI Premium* which is relatively expensive and outside the price range of many Power BI users.

Set up Cognitive Services in Azure

You must configure *Microsoft Cognitive Services* in *Azure* before you are able to leverage it in Power BI. You will have an *Azure account* if you went through the steps to create one in the book's introduction. If you are new to *Azure*, I highly recommend you sign up for a free account. At the time of writing, Microsoft was offering $200 in free credits for the first month! That is more than enough credits to cover all the *Azure*-related activities in this book if you were to do them in the first month of service. If you don't have an *Azure account,* go to the book's introduction to get the instructions needed to sign up for one.

If you already have an *Azure account,* then you need to sign into the *Azure Portal.* You can do so by going to `portal.azure.com`. After you have signed into your *Azure account,* click the *Create a resource* button at the upper left of the page. Next click *AI + Machine Learning* ➤ *Text Analytics.* Those actions will take you to a form that you need to fill out to set up the *Text Analytics* service. Give the resource a meaningful name in the *Name* textbox, attach a subscription in the *Subscription* textbox, set the location to a region closest to you in the *Location* textbox, choose a pricing tier that is appropriate for your needs in the *Pricing tier* textbox, and attach the resource to an existing resource group or create a new one in the *Resource group* textbox. If you are still in your first month, the subscription with the $200 credits will be available to you as an option for subscriptions.

The Data Science Virtual Machine (DSVM)

As stated earlier, this example will only be done in Python because R is not supported. In addition to using just Python, the example will be done in the *Microsoft Data Science Virtual Machine (DSVM).* The reason why the *DSVM* is recommended is because it is pre-configured with the majority of the tools you need for data science. It has the *Anaconda* distribution of Python pre-installed, *Power BI, SQL Server, Jupyter Lab*, and *Azure Data Studio,* to name a few. Configuring such an environment yourself would be hard to do, but with the *DSVM,* you can spin one up on demand.

Before you move forward, there are some things you need to do in your *DSVM* to finish configuring your environment. You need to install some packages for Microsoft Cognitive Services. Here are the steps you need to perform:

1. *Launch your DSVM*: You can do your development on your personal desktop, but I highly recommend that you use the *DSVM*. If you choose to go your own route, then start from your environment of choice.

2. *Open the conda prompt*: To open the conda prompt, search for it in the windows search bar by typing anaconda prompt. You should see it appear in your search list. Right-click it, then launch it as an administrator.

3. *Activate the Python environment you use for Azure development*:
   It is recommended that you use the *AzureML* environment
   that comes pre-installed in the *DSVM*. To activate the *AzureML*
   environment, type the following in the conda prompt:

   ```
   conda activate AzureML
   ```

   If you are operating in your own environment and don't know
   the environment you should use, type conda env list in the conda
   prompt to get a list of the environments available to you.

4. *Type the following code to install the library*: Make sure you are
   installing the library in the appropriate environment by activating
   the desired environment using the code in Step 3.

   ```
   pip install azure-cognitiveservices-language-textanalytics
   ```

Performing sentiment analysis in Microsoft Cognitive Services via Python

You will use the freely available data set of Yelp reviews from Kaggle in this exercise. Here are the steps you need to go through to acquire the data and score it in Power BI.

Step 1: Get the Yelp review data from Kaggle

At the time of the writing, the data set was available at this URL: www.kaggle.com/yelp-dataset/yelp-dataset/version/4. Make sure to download the *csv* version of the *yelp_review* data set for this example. The data set will be downloaded as a zip file. Unzip the *yelp_review.csv* file and save it to the *Data* folder for this chapter in the repository.

The *yelp_review.csv* data set is big. It contains close to 5.3 MM reviews. You do not need that many samples for this exercise, and it is *highly* recommended that you use a sample to keep cost down to a minimum. The following script gets a random sample of 100 reviews from the *yelp_review.csv* data set and saves it under the *yelp_review_sample.csv* file name:

```
import pandas as pd
import numpy as np
import os
```

```
os.chdir("set to your working directory")

dfYelpReviews = pd.read_csv("yelp_review.csv")
dfYelpReviewsSample = dfYelpReviews.sample(100, random_state = 1)
dfYelpReviewsSample["id"] = np.arange(len(dfYelpReviewsSample))
dfYelpReviewsSample = dfYelpReviewsSample[["id","text"]]
dfYelpReviewsSample.to_csv(
    "yelp_review_sample.csv", index = False)
```

Here's how the script works:

1. *Loads the required libraries*: *Pandas* is used for some basic data wrangling, the *arange* function from *numpy* is used to create a row number column, and *os* is used to set the working directory.

2. *Reads in the yelp_review data set*: This is accomplished by using the *read_csv()* method of the *pandas*.

3. *Creates a sample of the data set using the sample method of the dfYelpReviewsSample data frame*: You created the sample using just two arguments. The first argument is the sample size which, in this case, is 100. The second argument is the *random_state* argument. If you set the argument to 1 as earlier, then your sample will return the same data set as the one used by the book's author.

Step 2: Import the necessary libraries, modules, and function for the script

This step is where the development of the script begins. You will start with the imports shown here:

```
import pandas as pd
import ast
import os
from azure.cognitiveservices.language.textanalytics import
TextAnalyticsClient
from msrest.authentication import CognitiveServicesCredentials
```

You are aware of two of the first three libraries. The *ast* library is new. The *literal_eval* function from that library is used to evaluate a string that represents a python expression. That functionality is used in this script. It will be explained in a subsequent step. The *TextAnalyticsClient* and *CognitiveServicesCredentials* functions are used to communicate with *Microsoft Cognitive Services* in the next step.

Step 3: Assign values to the variables used in the script

The values of the variables are set in the script using the following code:

```
api_key = "<put api key here>"
endpoint = \
    "https://rwtextanalyticsapi.cognitiveservices.azure.com/"
credentials = CognitiveServicesCredentials(subscription_key)
text_analytics = TextAnalyticsClient(
    endpoint=endpoint, credentials=credentials)
```

You can find the value of your *API key* and *endpoint* by going to your text analytics resource in *Azure*. Both can be found in the *Overview* page. You can get there by clicking the *Overview* tab located in the upper right corner. The *API key* is used to create an instance of *CognitiveServicesCredentials*. The *credentials* and the *endpoint* variables are used together to create an instance of *TextAnalyticsClient* which is assigned to the *text_analytics* variable. The *text_analytics* variable will be the workhorse for doing the sentiment analysis.

Step 4: Read in the sample of the Yelp review data into Python

The following code reads in the sample into a *pandas* data frame. **It is important that you use the sample data set and not the entire data set.** The entire data set contains close to 5.3 MM reviews which could cause you to incur unnecessary charges. The 100 observations in the sample data set are more than enough for practice. The sample data set that you created earlier contains two fields: the *id* field is used to identify each review, and the *text* field contains the review that will be analyzed. The language field is added so that Microsoft Cognitive Services know what language to use. You are using *en* because that is the code for English which is the language the reviews are in.

```
df = pd.read_csv("< path to yelp_sample.csv>")
df["language"] = "en"
```

Step 5: Transform the data frame to the format that is required by Microsoft Cognitive Services

Microsoft Cognitive Services needs the data sent to it for sentiment analysis to be a list of dictionaries. Here is what a *list* of *dictionaries* looks like:

```
documents = [
                {"id": "1",
                 "language": "en",
                 "text": "Football is great sport."},
                {"id": "1",
                 "language": "en",
                 "text": "Lamar Jackson is the best quarterback."},
                {"id": "1",
                 "language": "en",
                 "text": "UK football sucks!!! "}
]
```

The *to_json* method of the *pandas* data frame can be used to reformat the *df* data frame into that format. In order to do so, you need to set the *orient* argument to *records* as done in the following code:

```
documents = df.to_json(orient='records')
```

The output is formatted as a *list* of *dictionaries,* but Python does not see it that way. Python just sees a string. You need to explicitly tell Python to evaluate the output of df. to_json(orient='records') as a python expression. You can accomplish that by using the *literal_eval()* function from the *ast* library as illustrated in the following code:

```
documents = ast.literal_eval(documents)
```

Now the documents variable contains the information *Microsoft Cognitive Services* needs in the proper format. Each *dictionary* in the *list* passed to *Microsoft Cognitive Services* contains three key/value pairs. The three keys are the *id* which is used to uniquely identify each review that you are performing sentiment analysis on, the *text* which represents the text you want to analyze, and the *language* you want to do your sentiment analysis in.

Step 6: Score the reviews using the sentiment method of the text_analytics object

The code used to score the reviews is as follows:

```
response = text_analytics.sentiment(documents=documents, raw = False)
```

It returns a *list* that contains a *batch result item* for each review that was passed to the *sentiment()* method. Each *batch result item* has properties that hold information about the review. The two properties that you are interested in are the *id* property and the *score* property. You will use those two properties to build a data frame that contains the sentiment scores in the next step.

Step 7: Create a data frame to hold the sentiment data

You were able to score your reviews and get information returned backed to you in the form of a list of *batch result items*. As stated earlier, the *list* contains *batch result items* for each review that was sent to *Microsoft Cognitive Services*. Each *batch result items* contains properties that expose information about the review. The two that you are interested in are the *id* of the review and the *sentiment score.*

Here is the section of the script that uses the *id* and *score* property of each object returned by *Microsoft Cognitive Services* to create the data frame:

```
listSentiments = []
for document in response.documents:
    id = document.id
    score = document.score
    listSentiments.append([id, score])

dfSentiments = pd.DataFrame(
    listSentiments, columns=['ID','Score'])

dfSentiments
```

Here is an explanation of the script broken out in steps:

1. *Create an empty list and name it listSentiments*: The first line creates an empty list named *listSentiments*. This list will get populated with the output retrieved from *Microsoft Cognitive Services* and will ultimately get converted to a data frame in a subsequent step.

2. *Iterate through each document in response.documents using the for loop*: The *for* loop is used to iterate through each *document* in *response.documents* to access the sentiment information.

3. *Grab the id and score for each document*: The *documents* returned by each iteration contain an *id* and *score* property. You grab those values and put them in their respective variables using the following code:

```
id = document.id
score = document.score
```

4. *Append the values from the current document to the listSentiments list*: In this step, you populate the *listSentiments* list with the variables created in Step 3 via the list's append method. The append method uses a list based on the two variables to append the information.

5. *Create the dfSentiments data frame*: The *DataFrame()* method from *pandas* is used to convert the *listSentiments* list to a data frame. A *list* containing the desired column names is passed to the *columns* argument of the *DataFrame()* method to name the columns in the data frame.

Step 8: Add the data to Power BI

The complete script is in Listing 9-3.

Listing 9-3. Python script that uses Microsoft Cognitive Services to perform sentiment analysis

```python
import pandas as pd
import json
import ast
from pandas.io.json import json_normalize
from azure.cognitiveservices.language.textanalytics import (
    TextAnalyticsClient)
from msrest.authentication import CognitiveServicesCredentials

api_key = "<put api key here>"
endpoint = \
    "https://rwtextanalyticsapi.cognitiveservices.azure.com/"
credentials = CognitiveServicesCredentials(api_key)
text_analytics = TextAnalyticsClient(
    endpoint=endpoint, credentials=credentials)

df = pd.read_csv("<path to yelp_sample.csv>")
df["language"] = "en"

documents = df.to_json(orient='records')
documents = ast.literal_eval(documents)
response = text_analytics.sentiment(
    documents=documents, raw = False)

listSentiments = []
for document in response.documents:
    id = document.id
    score = document.score
    listSentiments.append([id, score])

dfSentiments = pd.DataFrame(
    listSentiments, columns=['ID','Score'])
```

Copy and paste the code into the *Python editor* in Power BI. Doing so will expose the *dfSentiments* data frame to the Power BI data model.

Applying AI to your Power BI data model using services other than Microsoft Cognitive Services

Microsoft Cognitive Services has a nice selection of AI models that you can leverage, but they do not cover all situations. Fortunately, other vendors such as IBM and Google offer services that you can use that are extremely powerful that can help fill in the gap. You will find that some services are common among all vendors such as *sentiment analysis,* but there are other services that may be unique to a given vendor. One such service is the *Tone Analyzer* offered by *IBM Watson Natural Language Understanding*.

In this chapter, you will learn how to leverage *IBM Watson Natural Language Understanding* to take advantage of its *Tone Analyzer API* in Power BI. The *Tone Analyzer* is used to return the tone of the text. Some of the tones it can detect are happy, sad, and excited, to name a few. This exercise will be done in just Python because R was not supported by *IBM Watson* at the time of this writing. The next section will cover how to set up your account in *IBM Watson Natural Language Understanding,* and that will be followed by a section that covers how to use the service from a Python script in Power BI.

Configuring the Tone Analyzer service in IBM Watson

Like with Azure, you need to get an account, then set up the *Tone Analyzer* service. Here are the steps to do that.

Step 1: Sign up for IBM Cloud account

Go to `www.ibm.com/cloud/watson-natural-language-understanding/pricing` to sign up for an *IBM Watson Natural Language Understanding* account.

Step 2: Log into the IBM Cloud

After you sign up, go to this URL to log in: `https://cloud.ibm.com/login`.

Step 3: Go to the Tone Analyzer page

1. Click the *Create Resource* button at the top of the page. It will take you to a page that has a big list of services available to you.

2. Search for the *Tone Analyzer* service in the *Search the catalog...* search box contained in the upper right portion of the page. Doing so will filter the list down to the *Tone Analyzer* service. Click the *Tone Analyzer* service when you see it.

Step 4: Define your Tone Analyzer service

1. In the *Select a region* section, pick the region that is closest to you in the drop-down box.

2. Choose the *Lite* plan in the *Select a pricing plan* section. That plan is sufficient for this exercise. You may need to upgrade your plan if your future needs take you past the free 2500 API calls you receive per month or if you have privacy concerns.

3. In the *Configure your resource* section,

   • Give your plan a meaningful name in the *Service name:* section.

   • Select a resource group. If you are new to *IBM Watson Natural Language Understanding,* you can go with the default resource group. If you already have resource group(s) in *IBM Watson Natural Language Understanding,* they will be available for you to select. Like in Azure, resource groups allow you to combine your *IBM Watson Natural Language Understanding* resources in one group which facilitates access and billing.

   • You can optionally add tags in the *Tags:* section which further helps you organize and group your resources.

4. Click the *Create* button in the upper left to create the resource.

Step 5: Get your API key

Your API key will be used to authenticate you when you make API calls to the *IBM Watson Natural Language Understanding* service. It is *very* important that you keep the API key secure.

You should have been taken to a page where you can administer your *Tone Analyzer* resource after you clicked the *Create* button in Step 4. If for some reason you were not, you can get to it by clicking the navigation icon in the upper right corner of the page which is pictured in Figure 9-1.

Figure 9-1. *The IBM Navigation menu*

One of the options you will see is *Resource List*. Click it to go to the *Resource list* page. You should see your *Tone Analyzer* resource in the *Services* section. Click it to go to the page that will allow you to administer your *Tone Analyzer* resource.

After you arrived at the desired location, click the *Manage* tab in the upper right corner pictured in Figure 9-2.

Figure 9-2. *Link to get to the page where you can manage your Tone Analyzer service*

Once there, you can get your API key by clicking the copy icon next to the API key illustrated in Figure 9-3.

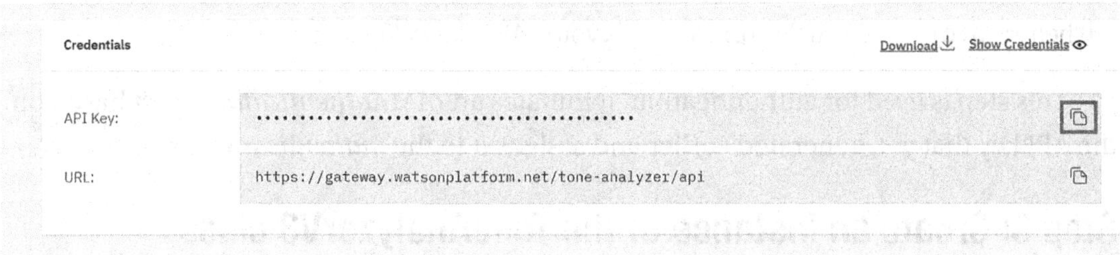

Figure 9-3. *Accessing your Tone Analyzer service API key*

Writing the Python script to perform the tone analysis

You can start developing the python script now that you have the resources set up in *IBM Watson Natural Language Understanding*. Just like in the *Microsoft Cognitive Services* example, the script will score the data and produce a data frame with the scored data that can be brought into Microsoft Power BI. Here are the steps.

Step 1: Import the required libraries and modules

You will start off by defining the imports you need for the script as shown in the following code:

```
import json
import pandas as pd
import ast
import os
from ibm_watson import ToneAnalyzerV3
from ibm_watson.tone_analyzer_v3 import ToneInput
from ibm_cloud_sdk_core.authenticators import IAMAuthenticator
```

This script uses the *pandas* library to handle some basic data wrangling tasks and json work. The *literal_eval* function from the *ast* library is used to execute a string representation of some python code, and the *ToneAnalyzerV3* and *IAMAuthenticator* are used to perform the tone analysis. More explanations about the latter two will be given in a subsequent step.

Step 2: Create an instance of the IAMAuthenticator class

The code needed to do so is as follows:

```
authenticator = IAMAuthenticator("<your API key>")
```

This step is used for authentication. It initiates an *IAMAuthenticator* object based on the API key that you generated earlier and assigns it to the *authenticator* variable.

Step 3: Create an instance of the ToneAnalyzerV3 class

The code needed to do so is as follows:

```
service = ToneAnalyzerV3(
                    version='2019-12-22',
                    authenticator=authenticator
        )
```

The code creates an instance of the *ToneAnalyzerV3* class that you will use to communicate with the *IBM Watson Natural Language Understanding* Tone Analyzer service. You need to populate two parameters: the *version* parameter and the *authenticator* parameter. The version parameter is used to tell the API which version of the API you want to use. If a version of the API was created on the date you specified, then that version of the API will be used. Otherwise, it will use the most recent version of the API created before the data specified. It is important to use the date that was tested during development to prevent your code from breaking.

Step 4: Set the service URL of the service object

The code needed to perform this step is as follows:

```
service.set_service_url(
    "https://gateway.watsonplatform.net/tone-analyzer/api")
```

You can find your service URL by going to the Manage tab in your *IBM Watson Natural Language Understanding* Tone Analyzer resource page.

Step 5: Create a data frame that will hold the data you want to use for the tone analysis

The code is straightforward. You use the read_csv() method from *pandas* to retrieve the data as illustrated here:

```
dfDocuments = pd.read_csv(
    "<path to the documents csv file>")
```

Step 6: Create a data frame based on the scored data

Use the following code to create an empty list:

```
listReturnedUtterance = []
```

The tone analysis will be done by sending each *document* to *IBM Watson Natural Language Understanding* individually, and the results will be sent back to you in JSON format.

A *document* in tone analysis is a group of words or sentences that represents the level of your analysis. Examples include a tweet if you are doing tone analysis for Twitter or a restaurant review if you are doing tone analysis on the restaurant reviews.

The returned JSON will be parsed to get the tones with their corresponding scores. That information will be stored in the list created in the line of code earlier. This list will be the basis of the data frame that will be the data source for Power BI.

Step 7: Create a looping structure to individually send the documents to IBM Watson

This is accomplished in the following code:

```
for index, row in dfDocuments.iterrows():
```

The *IBM Watson Natural Language Understanding* tone analysis service only allows 50 submissions in one batch, but you will often need to send more. You can get around this limitation by sending documents to *IBM Watson Natural Language Understanding* one request at a time instead of in batches. You can accomplish this by iterating over the *dfDocuments* data frame via the *iterrows()* method. The *iterrows()* method will iterate over the *dfDocuments* data frame and return a pair of index and row value. The *index* represents the index value of the row and the *row* is returned as a *pandas* series. This functionality enables you to act on each column of the data frame individually row by row.

Step 8: Format and score the document

The code required to perform this step is as follows:

```
submissionText = row["text"].replace("'","")
submissionText = submissionText[0:500]
PythonExpression = "[{'text':  '" + submissionText + "'}]"
Submission = ast.literal_eval(PythonExpression)
tone_chat = service.tone_chat(Submission).get_result()
```

IBM Watson Natural Language Understanding needs to receive the request in a form of a list with each element in the list being a dictionary that contains the document that needs to be analyzed. You are sending one request at a time so the request needs to be formatted as follows:

```
[{'text': '<document to be analyzed'}]
```

The result is a list that contains a dictionary with one element. Your data is in a data frame, so you need to do some manipulation to get the data in the preceding format.

In the first line of code in the preceding code snippet, you replace all *single quote* symbols in the *document* with an empty string. The single quote symbol is used to encapsulate the text in your document. Removing instances of them that appear in the document will prevent parsing errors.

The second line of code makes sure you are only getting the first 500 characters of the document because that is the max size allowed by IBM Watson. The 0 in that line tells Python what position you want to start in the string, and it is inclusive.

Python is a zero-based programming language, and R is a one-based programming language. That means the default initial index created in Python will be 0, but it will be 1 in R.

The 500 tells Python the position where you want to stop, but it is not inclusive. So, the preceding code will retrieve all characters starting in position 0 up to the position right before position 500 or the length of the string if the length is lower than 500. This results in a max of 500 characters being returned. This is the max number of characters allowed by IBM Watson.

The third line of code creates a string that represents the expression needed to create a list with a dictionary in it that contains the document you want to score. It is built by concatenating the static portions that will be the same for all submissions with the dynamic portion which contains the *document* to be analyzed.

The fourth line of code executes the string as a Python expression. The result of that, in this case, is a *list* object with a *dictionary* inside being produced. The resulting *list* is assigned to the *Submission* variable.

Lastly, the fifth line of code uses the *tone_chat()* method from the *service* object you created earlier to communicate the submission to *IBM Watson Natural Language Understanding*. The results are stored in the *tone_chat* variable.

Step 9: Assign the results of the tone analysis and set the initial values of the tone variables

The following code is used to perform this step:

```
utterances = tone_chat["utterances_tone"]

sad, frustrated, satisfied, excited, \
polite, impolite, sympathetic, unknown = \
    0.0, 0.0, 0.0, 0.0, 0.0, 0.0, 0.0, 0.0
```

The preceding first line gets the *utterances_tone* dictionary returned from *IBM Watson Natural Language Understanding* and assigns it to the *utterances* variable. The output of the *utterances_tone* comes back looking like the following sample JSON:

```
{'utterances_tone':
    [{'utterance_id': 0,
      'utterance_text': '<document that was analyzed>',
      'tones': [{'score': <between 0 and 1 for tone 1>,
                 'tone_id': '<tone_id 1>',
                 'tone_name': '<tone_name 1>'},

                 .........................................................................

                 {'score': <between 0 and 1 for tone n>,
                  'tone_id': '<tone_id n>',
                  'tone_name': '<tone_name n>'}
                ]
      }
    ]
}
```

The main part of the preceding JSON you are interested in are the tones. Note that multiple tones can be returned per *document* submitted. The possible tones are represented in the tone variables that you created in this step. Each possible tone was initialized with a value of 0.

Note that a \ was used for line continuation when initializing the tone variables. The three lines actually represent one line of code, but that line is too wide to fit on this page. The line continuations were used to overcome the width limitation.

For each tone returned, you get the score, an id, and the tone name. In the next step, you will iterate through the tones to get the score for each tone returned.

Step 10: Loop through the returned tones and assign their scores to the appropriate variable

The following code is used to accomplish this step:

```python
tones = utterances[0]["tones"]
for tone in tones:
    toneid = tone["tone_id"]
    if toneid == "sad":
        sad = tone["score"]
    elif toneid == "frustrated":
        frustrated = tone["score"]
    elif toneid == "satisfied":
        satisfied = tone["score"]
    elif toneid == "excited":
        excited = tone["score"]
    elif toneid == "polite":
        polite = tone["score"]
    elif toneid == "impolite":
        impolite = tone["score"]
    elif toneid == "sympathetic":
        sympathetic = tone["score"]
    else:
        unknown = tone["score"]

listReturnedUtterance.append(
    [index, sad, frustrated, satisfied, excited,
    polite, impolite, sympathetic, unknown])
```

Because you are submitting one document at a time, the returned value represents the tones for one submission. Even though tones for only one document are being returned, you still need to explicitly tell Python that you want the first one. You do that with the following code: utterances[0]["tones"]. The [0] gets the first utterance, and the ["tones"] extracts the tones from that utterance. The result is stored in the *tones* variable. Next, you iterate over the elements in *tones* variable and assign the score to the appropriate tone. So, if the value of *toneid* of the current iteration is *impolite,* then the score for the *impolite* variable will be changed from 0 to the impolite score. After all the returned tones have been iterated through and assigned to the appropriate tone variable, the index value of the current iteration along with the values of each tone is appended to the *listReturnedUtterance* list.

Step 11: Create a data frame based on the listReturnedUtterance list

The code needed to accomplish this step is as follows:

```
colnames = ["index", "sad", "frustrated", "satisfied", "excited",
"polite", "impolite", "sympathetic", "unknown"]

dfReturnedUtterance = pd.DataFrame(
    listReturnedUtterance, columns=colnames)
```

It is relatively easy to convert your *listReturnedUtterance* list to a data frame using the *DataFrame* method from *pandas*. You need to give it two arguments. The first argument is the list you want to convert into a data frame, which in this case is the *listReturnedUtterance* list. The second argument you need to populate is the *columns* argument. You supply that argument with a Python list that contains the names you want to use for the columns.

Step 12: Merge the dfReturnedUtterance data frame with the dfDocuments data frame

Here is the code needed to perform this step:

```
dfOutput = pd.merge(
    dfDocuments, dfReturnedUtterance,
    how='inner', left_on = 'id', right_on = 'index')

dfOutput = dfOutput[
    ['id', 'text', 'sad', 'frustrated', 'satisfied',
    'excited', 'polite', 'impolite', 'sympathetic', 'unknown']]
```

You merge the *dfReturnedUtterance* data frame to the *dfDocuments* data frame using the *merge* method of *pandas*. The way this is accomplished is you provide the data frame that represents the left side of the join as the first argument of the *merge()* method and the data frame that represents the right side of the join as the second argument. Next, you tell the *pandas* what type of join you want to do using the *how* argument. In this example, you are performing an *inner* join. Lastly, you identify the columns you want to use in the join from each data frame. The column that you want to use from the left data frame is specified in the *left_on* argument, and the column you want to use from the right data frame is specified in the *right_on* argument.

Step 13: Copy the complete script into Power BI

Here is the complete script that you need to copy and place in the *Python script editor* that you access via *GetData*. The script returns multiple data frames that will be exposed to Power BI. The one you want to add is the *dfOutput* data frame.

Listing 9-4. Python script that uses IBM Watson to perform tone analysis

```
import json
import pandas as pd
import ast
import os
from ibm_watson import ToneAnalyzerV3
from ibm_watson.tone_analyzer_v3 import ToneInput
from ibm_cloud_sdk_core.authenticators import IAMAuthenticator

authenticator = IAMAuthenticator("<your API key>")

service = ToneAnalyzerV3(
    version='2019-12-22',
    authenticator=authenticator)

service.set_service_url(
    "https://gateway.watsonplatform.net/tone-analyzer/api")

dfDocuments = pd.read_csv(
    "<path to the documents csv file>")
```

```python
listReturnedUtterance = []
for index, row in dfDocuments.iterrows():
    submissionText = row["text"].replace("'","")
    submissionText = submissionText[0:500]
    PythonExpression = "[{'text':   '" + submissionText + "'}]"
    Submission = ast.literal_eval(PythonExpression)
    tone_chat = service.tone_chat(Submission).get_result()

    utterances = tone_chat["utterances_tone"]

    sad, frustrated, satisfied, excited, \
    polite, impolite, sympathetic, unknown = \
        0.0, 0.0, 0.0, 0.0, 0.0, 0.0, 0.0, 0.0

    tones = utterances[0]["tones"]
    for tone in tones:
        toneid = tone["tone_id"]
        if toneid == "sad":
            sad = tone["score"]
        elif toneid == "frustrated":
            frustrated = tone["score"]
        elif toneid == "satisfied":
            satisfied = tone["score"]
        elif toneid == "excited":
            excited = tone["score"]
        elif toneid == "polite":
            polite = tone["score"]
        elif toneid == "impolite":
            impolite = tone["score"]
        elif toneid == "sympathetic":
            sympathetic = tone["score"]
        else:
            unknown = tone["score"]

    listReturnedUtterance.append(
        [index, sad, frustrated, satisfied, excited,
        polite, impolite, sympathetic, unknown])
```

```
colnames = ["index", "sad", "frustrated", "satisfied", "excited",
"polite", "impolite", "sympathetic", "unknown"]
dfReturnedUtterance = pd.DataFrame(
    listReturnedUtterance, columns=colnames)

dfOutput = pd.merge(
    dfDocuments, dfReturnedUtterance,
    how='inner', left_on = 'id', right_on = 'index')

dfOutput = dfOutput[
    ['id', 'text', 'sad', 'frustrated', 'satisfied',
    'excited', 'polite', 'impolite', 'sympathetic', 'unknown']]
```

In this chapter, we covered the technical details that you need to know to apply AI to your Power BI data models using custom models built in R and Python. You also learned how to easily perform data transformation required by your machine learning models that would be very hard to do in Power Query. Lastly, we covered how to use pre-built models from cloud-based services such as *Microsoft Cognitive Services* and *IBM Watson Natural Language Understanding*. The techniques in this chapter were self-service-based techniques and are not meant to be an enterprise solution. In Chapter 10, you will learn how to use similar techniques on an enterprise scale!

## CHAPTER 10

Productionizing Data Science Models and Data Wrangling Scripts

The data wrangling scripts and data scoring scripts in the previous chapters work great in a self-service situation or in small shops where one person is responsible for maintaining the Power BI data models. That is because in those situations you can get away with using the personal version of the *on-premises data gateway*. But, the enterprise version of the *on-premises data gateway* is required for enterprise solutions, and it does not allow the use of R or Python scripts embedded in Power BI. Fortunately, you can overcome this limitation using a relatively new feature in SQL Server known as *SQL Server Machine Learning Services (SSMLS)*. *SSMLS* is a feature of SQL Server that enables you to perform advanced data analytics inside the database via R and Python scripts that are wrapped in a special T-SQL stored procedure. Since you are able to fetch data via a stored procedure call using the *on-premises data gateway*, you can refactor your previously written data wrangling and data scoring scripts in Power BI to an enterprise solution by wrapping the scripts in a stored procedure.

In this chapter, we will go over some examples that show how to refactor a data wrangling script and a data scoring script, and as a bonus, we will show how to use a pre-built model that comes with *SSMLS* to get sentiment scores for free! First, let's refactor the Boston house price prediction example in *SSMLS*.

© Ryan Wade 2020
R. Wade, *Advanced Analytics in Power BI with R and Python*, https://doi.org/10.1007/978-1-4842-5829-3_10

Predicting home values in Power BI using R in SQL Server Machine Learning Services

In this example, we will refactor the R script that was used to predict house prices in Chapter 9 and wrap it in a special stored procedure. Doing so will enable the R script to be used with the enterprise version of the *on-premises data gateway*. This change will enable you to use the solution in an enterprise situation. Here are the required steps.

Build the R script that adds the model to SQL Server

Listing 10-1 shows the code needed to add the R model to SQL Server.

Listing 10-1. Code needed to add R model to SQL Server

```
library(RODBC)

# Load model into our R session
model <- readRDS("./Models/Model.rds")

# Connects to the database
server.name = "DSVM2019"
db.name = "BostonHousingData"
connection.string = paste(
    "driver={SQL Server}", ";",
    "server=", server.name, ";",
    "database=", db.name, ";",
    "trusted_connection=true", sep = "")
conn <- odbcDriverConnect(connection.string)

# Define parameters
model_name <- "R Model"
modelbin <- serialize(model, NULL)
modelbinstr = paste(modelbin, collapse = "")

# Build the SQL Statement and execute it
# using the sqlQuery command
sql_code <- paste0(
    "EXEC AddModel_R ",
```

```
    "@ModelName='",
    model_name, "', ",
    "@Model_Serialized='",
    modelbinstr, "'")
sqlQuery(conn, sql_code)

odbcClose(conn)
```

Now let's go over the code in steps.

Step 1: Load the necessary packages

```
library(RODBC)
```

Only one package needs to be loaded in this script which is the *RODBC* package. The *RODBC* package will be used to interact with SQL Server.

Step 2: Load the model into the R session

The following code is used to perform the task:

```
model <- readRDS("./Models/Model.rds")
```

In this step, you load the model into R using the *readRDS()* function which is a part of base R. You will be able to use a relative file path if your working directory is set to the R folder for Chapter 10.

Step 3: Create a connection to the database

The following code is used to connect to the database:

```
server.name = "DSVM2019"
db.name = "BostonHousingData"
connection.string = paste(
    "driver={SQL Server}", ";",
    "server=", server.name, ";",
    "database=", db.name, ";",
    "trusted_connection=true", sep = "")
conn <- odbcDriverConnect(connection.string)
```

A connection string is created based on the server and database name you used to set the server.name and db.name variables, respectively. That information is concatenated with some static information to build the connection string. The results are assigned to the *connection.string* variable. Lastly, the *connection.string* variable is passed to the *odbcDriverConnection()* function to create a connection object. The connection object that is created is named *conn*.

Step 4: Define the model variables

The code for this step is as follows:

```
model_name <- "R Model"
modelbin <- serialize(model, NULL)
modelbinstr = paste(modelbin, collapse = "")
```

A *model_name* variable is used to hold the name of the model. It will be used later in a subsequent step to identify the model in the *dbo.Models* table. The model is serialized using the *serialized()* function, and the results of the serialization is stored in the *modelbin* variable. Next, the *modelbin* is converted to a scalar character vector by using the *collapse* option of the *paste()* function, and the results are stored in the *modelbinstr* variable. That *modelbinstr* variable is the representation of the model that will be inserted into the database.

Step 5: Build the T-SQL statement to add the model to the database

The code to perform this task is as follows:

```
sql_code <- paste0(
    "EXEC AddModel_R ",
    "@ModelName='",
    model_name, "', ",
    "@Model_Serialized='",
    modelbinstr, "'")
```

The *paste0()* function is used to build the T-SQL that executes the *AddModel_R* stored procedure in SQL Server. The *AddModel_R* stored procedure uses the *INSERT* statement to insert the model into SQL Server using parameters that are based on the variables built in Step 4. The *VALUES* clause is used with the *INSERT* statement to pass the model name that is defined in the *model_name* variable and the string representation of the model that is defined in the *modelbinstr* variable. The *AddModel* stored procedure is already configured in the database that you will add to your server in a subsequent step. After you add the database, you can view the T-SQL code that is being executed by inspecting the code inside the dbo.AddModel_R stored procedure.

Step 6: Add the code needed to execute the T-SQL statement from R

Here's the code for this step:

```
sqlQuery(conn, sql_code)
odbcClose(conn)
```

The T-SQL code built in Step 5 is executed using the *sqlQuery()* function from the *RODBC* package. It uses the *conn* object to connect to the database and *sql_code* variable to give the function the T-SQL statement to execute. The connection to the database is closed after the T-SQL code is executed.

Step 7: Save the script

Make sure to save the script if you are starting from scratch. You will need to execute it in a later step to add the model to the SQL Server. Note that the script is available in the code repository of this book.

Use SQL Server Machine Learning Services with R to score the data

In this section, we will go over the steps needed to add the database used in this example to your server. You will also go over the steps needed to configure the database and score the data.

Step 1: Launch SQL Server Management Studio

If you are using the recommended *Windows 2019 DSVM,* then *SQL Management Studio (SSMS)* and *SQL Server 2019* will already be installed with *SQL Server Machine Learning Services* enabled. Please refer to the installation instructions located in the introduction if those resources are not installed in your environment.

Step 2: Create a connection to the server you want to use

When you launch *SQL Server Management Studio (SSMS)*, the server name should be populated in the textbox field. If it is not listed and you don't know the name of the server, then you can type "." and SSMS will open the local server in your environment.

Step 3: Add the BostonHousingInfo database to your server

A starter database for this exercise is in the *Databases* folder for this chapter. The name of the file that you will use to restore the database is *BostonHousingData.bak*. Perform the following steps to restore the *BostonHousingData* database:

1. Copy the *BostonHousingData.bak* file to the *Backup* folder for SQL Server. The location in the *DSVM* is C:\Program Files\Microsoft SQL Server\MSSQL15.MSSQLSERVER\MSSQL\Backup.

2. Go to *SSMS* and right-click the *Databases* folder, then select *Restore Databases....*

3. The *Restore Database* pop-up form will appear. Make sure you are on the *General* page tab. In the *Source* section, select the *Device* radio button.

4. Click the eclipse button, the button with the three dots, then click the *Add* button. Next, browse to the location where *BostonHousingData.bak* is located.

5. Select the *BostonHousingData.bak* file, then click the *OK* button to close out the *Locate Backup File* form.

6. Click *OK* to close out the *Select backup devices* form.

7. Click *OK* to close out the *Restore Database* form. Doing so will add the database.

The database added in this step will include the necessary infrastructure needed for this exercise and the data that will be scored. The *Boston Housing* data used in this example was obtained using the *load_boston()* function from *sklearn.datasets*. The code used to acquire the data set is in the code repository.

Step 4: Add the model to the database

Go back to *AddModelToDatabase.R* script in R Studio. Make sure that the working directory is set to the *R* folder. Execute the script. If you configured the script correctly, the R model will be added to the *dbo.Models* table in the *BostonHousingInfo* database.

Step 5: Add the stored procedure to the database that will do the scoring

Listing 10-2 is the code that creates the stored procedure.

Listing 10-2. The uspPredictHousePrices_R stored procedure

```
CREATE PROCEDURE [dbo].[uspPredictHousePrices_R]

AS

BEGIN

  -- Define variables
  DECLARE @model varbinary(max) =
      (SELECT MODEL
        FROM [dbo].[Models]
        WHERE ModelName = 'R Model');
  DECLARE @RScript nvarchar(max);
  DECLARE @Query nvarchar(max);
  DECLARE @InputDFName nvarchar(25);
  DECLARE @OutputDFName nvarchar(25);

  -- Define source data
  SET @Query='SELECT [crim], [rm], [tax], [lstat]
              FROM [dbo].[BostonHousingInfo]'

  -- R script to score data
  SET @RScript = N'
bhmodel_deserialized <-
    unserialize(as.raw(bhmodel_serialized));
model_data <- dfInputData
pred_medv <-
    predict(bhmodel_deserialized, model_data)
```

```
dfOutputData <-
    cbind(model_data, pred_medv)'

  SET @InputDFName = 'dfInputData'
  SET @OutputDFName = 'dfOutputData'

  EXEC sp_execute_external_script
      @language = N'R'
      ,@script = @RScript
      ,@input_data_1 = @Query
      ,@input_data_1_name = @InputDFName
      ,@output_data_1_name = @OutputDFName
      ,@params = N'@bhmodel_serialized varbinary(max)'
      ,@bhmodel_serialized = @model

  WITH RESULT SETS((
      [crim] float
      ,[rm] float
      ,[tax] float
      ,[lstat] float
      ,[pred_medv] float
  ));

END
```

The preceding code performs the following actions:

1. Defines the following variables needed in the script:

    - *@model* variable is used to hold the model. You retrieve the model from the dbo.Models table using T-SQL.

    - *@RScript* variable holds the R script that does the scoring.

    - *@query_string* variable holds the TSQL script that defines the data set that will be scored.

    - *@InputDFName* variable holds the name of the input data set that Python will use to refer to the input data set.

    - *@OutputDFName* variable holds the name of the output data set that R will use to refer to the output data set.

2. Sets the *@Query* with the TSQL code that returns the input data set. The input data set represents the data that will be scored using R.

3. Sets the *@RScript* variable with the R code that will perform the scoring. The R script assigned to this variable performs the following steps:

   - Loads the *bhmodel_serialized* in a raw format, then unserializes it using the *unserialize()* function and assigns the results to the *bhmodel_deserialized* variable.

   - Create a variable named *model_data* that is based on the *dfInputData* data frame. You do this because it is best practice to not act directly on the input data set passed to R from SQL Server.

   - Uses the *predict()* function from base R to score the data. The first parameter passed to the function is the model, and the second parameter is the data that needs to be scored. The output is a vector that contains the predictions, and it will be stored in the *pred_medv* variable.

Note that you were able to include the whole data frame. That is because it only contains the columns needed by the regression formula. In more advanced situations, you may need to perform transformation to the data passed to R to get it in the structure that is needed by the model's formula.

   - Creates a *dfOutputData* data frame using the *cbind()* function to do a column-based bind that combines the *model_data* data frame and the *pred_medv* vector into one data frame.

   - Sets the names you want to use for the input and output data frames.

4. Configures the *sp_execute_external_script* stored procedure.
    Most of this is self-explanatory, but the *@params* and
    @bhmodel_serialized parameters may need some explanation.
    The *@params* argument enables you to define extra parameters
    needed by the model. In this case, we are defining a parameter
    named *@bhmodel_serialized* with a *varbinary(max)* data type.
    This parameter will pass the model to the R script and will
    expose it to the script with the parameter name minus the @ sign
    which happens to be in this case *bhmodel_serialized*. The value
    of @bhmodel_serialized parameter is set in the succeeding line.

5. The *WITH RESULT SETS* clause is used to define the column
    names and column data types. This will give Power BI the
    information it needs to determine the column names and data
    types of the scored data set passed to it from SQL Server.

Execute the preceding script to add the stored procedure to the database. Now, you
will be able to score the data by calling the stored procedure.

Step 6: Fetch the scored data in Power BI from SQL Server

To fetch the scored data in Power BI, launch Power BI, then go to *GetData* ➤ *SQL
Server*. Populate the *Server and Database (optional)* textbox. Next, expand the *Advanced
options* section and type EXEC [dbo].[uspPredictHousePrices_R] in the *SQL statement
(optional, requires database)* textbox. The *Include relationship columns* check box is
checked by default, but it is not needed in this situation, so it is unchecked. Figure 10-1
shows what the form looks like after it has been configured.

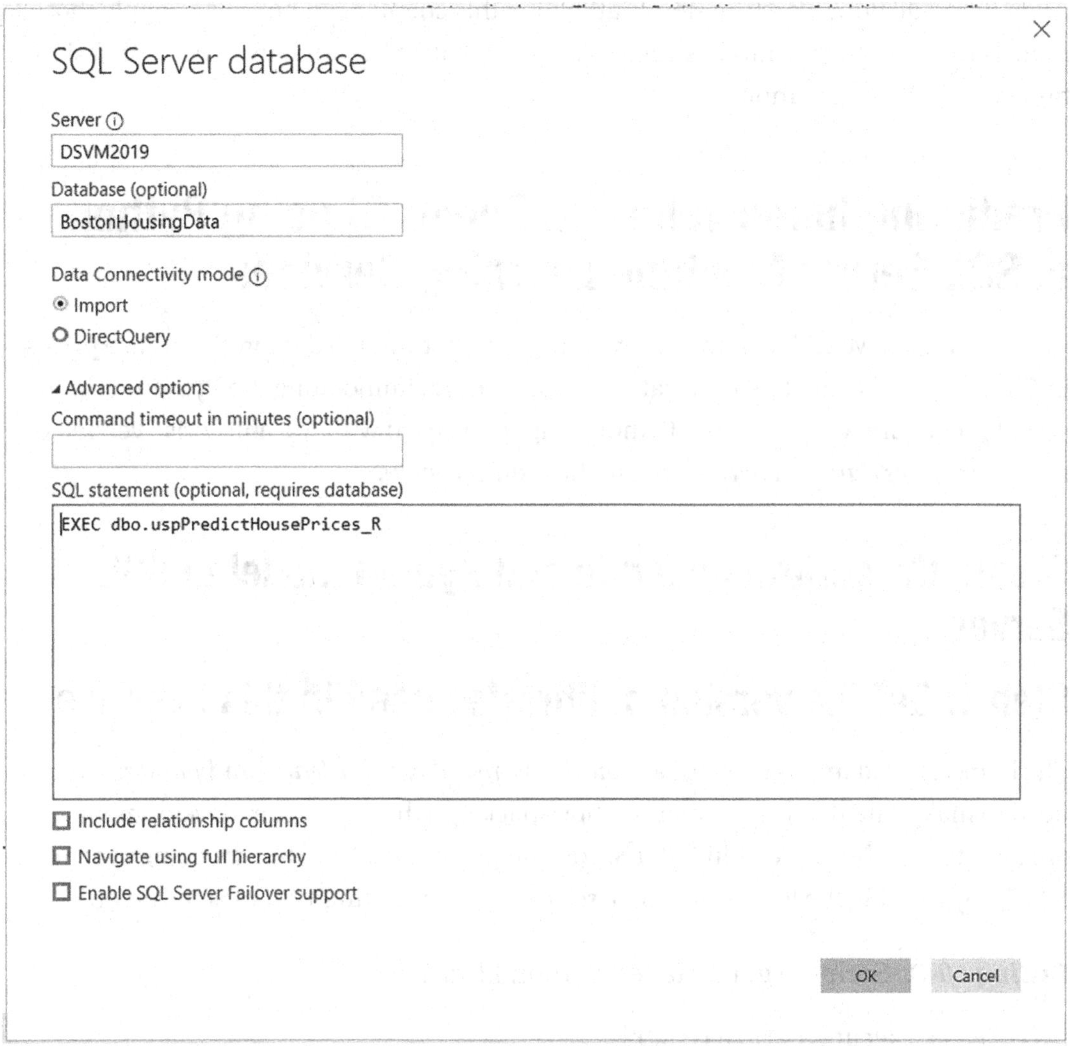

Figure 10-1. *GetData form populated with information needed to execute the dbo.uspPredictedHousePrices_R stored procedure*

Click *OK* to add the data to the Power BI data model.

The focused in this example should not be on the model. The model in this exercise was based on a simple linear regression formula, and it was made simple so that the focus would be on how to implement an R model to score data in Power BI via *SSMLS*. We did not focus on developing a complex model on purpose. Developing accurate machine learning models is well beyond the scope of this book. But if you or the data scientist on your team already knows how to develop complex models in R that produce accurate predictions, then the steps outlined in this chapter can be used to apply those R models to Power BI data models via *SSMLS*. Next, you will learn how to perform a similar task in *SSMLS* using Python.

Predicting home values in Power BI using Python in SQL Server Machine Learning Services

In this example, we will refactor the Python script that was used to predict house prices in Chapter 9 and wrap it in a special stored procedure. Embedding the Python script in a stored procedure will enable the Python script to be used with the enterprise version of the *on-premises data gateway*. Here are the required steps.

Create the script needed to add Python model to SQL Server

Step 1: Get the version of libraries used in this exercise

The libraries used in this exercise are *pickle, os, pyodbc, scikit-learn*, and *pandas*. You need to make sure that the versions of those packages that you use in development are the same as the versions in *SSMLS*. You can get the versions that are being used in *SSMLS* by running the following T-SQL script shown in Listing 10-3 in SQL Server.

Listing 10-3. Script to get installed Python libraries

```
EXECUTE sp_execute_external_script
    @language = N'Python'
    ,@script = N'
import pkg_resources
import pandas as pd
installed_packages = pkg_resources.working_set
installed_packages_list =
  sorted(["%s==%s" % (i.key, i.version) for i in installed_packages])

df = pd.DataFrame(installed_packages_list)
OutputDataSet = df'

WITH RESULT SETS (( PackageVersion nvarchar (150) ))
```

The preceding script returns a list of all of the packages that are installed in the installation of Python in *SSMLS*. As of the writing, here are the versions of the libraries used in *SSMLS 2019*:

- pyodbc 4.0.25

- scikit-learn 0.20.2

- pandas 0.23.4

Note that the *os* and *pickle* libraries are not listed. That is because those packages are included in all python installations by default.

Step 2: Create a conda environment

We need to make sure that all development is done using the same version that is used in *SSMLS* to prevent unnecessary version conflicts, to ensure that we need to set up a special development environment known as a *conda* environment.

---

conda environment helps you isolate project development by letting you determine the version of python you want to use, the libraries you want to include, and the version of those libraries that you want to use. There are other environment management options in python. What makes *conda* different from the others is that it is language agnostic so it can be used with other programming languages.

Perform the following steps to create a *conda environment*:

1. **Open VS Code**

   Open the command prompt and type the following code to change the working directory:

   ```
   cd <"path to Python folder for Chapter One">
   ```

   Next, type the following code to open VS Code from the preceding location:

   ```
   code .
   ```

2. **Open a terminal shell in VS Code in the above location**

Type *CTRL+SHIFT+`* to open a terminal shell. You are using the
Command Prompt in this example. If you need to change the
terminal shell, open the shell option window located in the upper
right pictured in Figure 10-2.

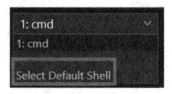

Figure 10-2. *The VS Code shell option window*

Once you make the preceding selection, you will be given the
following options depicted in Figure 10-3.

Figure 10-3. *Terminal shell options*

Make the highlighted selection to change the terminal shell to the
Command Prompt.

3. **Create a new conda environment called *ssmls***

We need a development environment that uses the same version
of Python and the same version of the libraries that are used
in *SSMLS*. The version of Python that is used in *SSMLS 2019* is
Python 3.73. Type the following code in the command prompt to
create an environment using that version of Python:

```
conda create --name ssmls python=3.7.3
```

4. **Switch to the environment you just created**

    In Step 3, you created an environment, but you are not working out of it. If you are working in the recommended *Data Science Virtual Machine,* then you will have many pre-installed environments as well as any environments you created. You need to type the following code in the command prompt to get a list of the environments that are available to you:

    ```
    conda env list
    ```

    If you are working in the *DSVM*, you will see several environments plus the *ssmls* environment. If you are working on your personal machine, then you may just see the *ssmls* environment, the base environment, and any other environments you may have created personally. Type the following command in the command prompt to activate the *ssmls* environment:

    ```
    activate ssmls
    ```

    Now you are in the *ssmls* environment which is based on the same version of Python that is used in *SSMLS 2019.*

5. **Install the required libraries in the ssmls environment**

    Now you need to install the *pandas, scikit-learn,* and *pyodbc* libraries. Let's install *pandas* first. To install the version of pandas needed for this task, type the following code:

    ```
    conda install pandas=0.23.4
    ```

    The preceding code will begin the installation process. You will have to go through several prompts to complete the installation. Perform the same action for *scikit-learn 0.20.2* and *pyodbc 4.0.25.* Make sure to put an = sign in between the library name and the version as illustrated in the *pandas* example.

Step 3: Create the code that pushes the model to SQL Server

The code in Listing 10-4 is used to push the model to SQL Server.

Listing 10-4. Script to push the model to the database

```
import os
import pyodbc
import pickle

server = 'DSVM2019'
database = 'BostonHousingData'
con = pyodbc.connect(
    'Trusted_Connection=yes',
    driver = '{SQL Server}',
    server = server,
    database = database
    )

cursor = con.cursor()

model = pickle.load(open("./Models/model.pkl", "rb"))
modelstr = pickle.dumps(model)

cursor.execute(
    "INSERT INTO [dbo].[Models](ModelName, Model) VALUES (?, ?)",
    "Python Model", modelstr
    )

con.commit()
con.close()
```

The preceding code uses the *os* library to interact with the file system, the *pyodbc* library to interact with SQL Server, and the *pickle* library to load and save the model object. Here are the steps that are performed in the script:

1. A connection object named *con* is created based on a *Windows Authentication* type connection. The connection gets the server name and database name from the server and database variables that were defined earlier

2. A cursor object named *cursor* is created that will be used to execute a T-SQL statement.

3. The model that was previously saved to disk is loaded into the Python session using the *open()* function and the *pickle.load()* function. The *open()* function opens the model using the "rb" access mode where "rb" stands for read binary. The results are then loaded to the session using the *pickle.load()* function which is assigned to a variable named *model*. Next, the model in the *model* variable is serialized using the *pickle.dumps()* function, and the result is assigned to the *modelstr* variable.

4. The *execute* method of the cursor object is used to execute a T-SQL statement on the targeted database. The first argument is the T-SQL statement. The TSQL statement is an INSERT statement that is used to insert the model and model name in the designated table. The ? symbols are used as place holders for the *ModelName* and *Model* parameters. The next two arguments are used to provide the parameters. Please note that order is important.

5. Save the script under the name *AddModelToDatabase.py*. Keep this file open. We will be coming back to it in a later step.

Use SQL Server Machine Learning Services with Python from Power BI to score the data

In this section, you will add the *BostonHousingData* database to your server and the script you created in the previous step to add the model to that database, then create a Python stored procedure to predict house prices. What follows are the steps to do so.

Step 1: Launch SQL Server Management Studio

If you are using the recommended *Windows 2019 DSVM,* then *SQL Management Studio (SSMS)* and *SQL Server 2019* will already be installed with *SSMLS* enabled. I highly recommend using the *DSVM* because it greatly reduces the effort required to configure your environment. If you are not using the *DSVM*, you will need to manually install *SQL Server 2019* with *SSMLS* enabled. You will also need to install *SQL Server Management Studio*. Installation instructions are included in the *Introduction*.

Step 2: Create a connection to the server you want to use

When you launch *SQL Server Management Studio,* the server name should be populated in the textbox field. If it is not listed and you don't know the name of the server, then you can type "." and *SSMS* will open the database server that is local to your environment.

Step 3: Add the BostonHousingInfo database to your server

A starter database for this project is in the database folder for this chapter. The name of the file is *BostonHousingData.bak.* If you did the R version of this exercise, you can skip this step because the database will already be added to the server. Otherwise, perform the following steps to restore the database:

1. Copy the *BostonHousingData.bak* file to the *Backup* folder for SQL Server. The location in the *DSVM* is C:\Program Files\Microsoft SQL Server\MSSQL15.MSSQLSERVER\MSSQL\Backup.

2. Go to *SQL Server Management Studio (SSMS)* and right-click the *Databases* folder, then select *Restore Databases....*

3. The *Restore Database* pop-up form will appear. Make sure you are on the *General* page tab. In the *Source* section, select the *Device* radio button.

4. Click the eclipse button, the button with the three dots, then click the *Add* button. Browse to the location where the *BostonHousingData.bak* is located.

5. Select the *BostonHousingData.bak* file, then click the *OK* button to close out the *Locate Backup File* form.

6. Click *OK* to close out the *Select backup devices* form.

7. Click *OK* to close out the *Restore Database* form.

The database added in this step will include the necessary infrastructure needed for this exercise and the data that will be scored. The *Boston Housing* data was obtained using the *load_bostondata()* function from *sklearn.datasets.* A script to get the data is included in the book's code repository.

Step 4: Add the model to the database

Go back to *AddModelToDatabase.py* script in VS Code. Make sure that the working directory is set to the *Python* folder. Execute the script. If you configured the script correctly, the model will be added to the *dbo.Models* table in the *BostonHousingInfo* database.

Step 5: Add the stored procedure to the database that will do the scoring

Listing 10-5 is the code that creates the stored procedure.

Listing 10-5. The uspPredictHousePrices_Python stored procedure

```
CREATE PROCEDURE [dbo].[uspPredictHousePrices_Python]
AS
BEGIN

  DECLARE @model VARBINARY(max) =
      (SELECT MODEL
        FROM [dbo].[Models]
        WHERE ModelName = 'Python Model');
  DECLARE @PythonScript nvarchar(max);
  DECLARE @Query nvarchar(max);
  DECLARE @InputDFName nvarchar(25);
  DECLARE @OutputDFName nvarchar(25);

  -- Define source data
  SET @Query='SELECT [crim], [rm], [tax], [lstat]
              FROM [dbo].[BostonHousingInfo]'

  -- Python script to score data
  SET @PythonScript = N'
import pickle
from sklearn import linear_model

bhmodel_deserialized = pickle.loads(bhmodel_serialized)
pred_medv = bhmodel_deserialized.predict(dfInputData)
dfOutputData = dfInputData
dfOutputData["pred_medv"] = pred_medv
'
```

```
    SET @InputDFName = 'dfInputData'
    SET @OutputDFName = 'dfOutputData'

    EXECUTE sp_execute_external_script
            @language = N'Python'
           ,@script = @PythonScript
           ,@input_data_1 = @Query
           ,@input_data_1_name = @InputDFName
           ,@output_data_1_name = @OutputDFName
           ,@params = N'@bhmodel_serialized varbinary(max)'
           ,@bhmodel_serialized = @model
  WITH RESULT SETS((
           [crim] float
          ,[rm] float
          ,[tax] float
          ,[lstat] float
          ,[pred_medv] float
    ));
END;
```

The preceding code performs the following actions:

1. Defines the following variables needed in the script:

    - *@model* variable is used to hold the model. The Python model that is assigned to this variable is retrieved from the dbo.Models table using T-SQL.

    - *@PythonScript* variable holds the Python script that does the scoring.

    - *@Query* variable holds the TSQL script that defines the data set that will be scored.

    - *@InputDFName* variable holds the name of the input data set that Python will use to refer to the input data set.

    - *@OutputDFName* variable holds the name of the output data set that Python will use to refer to the output data set.

2. Sets the *@Query* with the TSQL code that returns the input data set.

3. Sets the *@PythonScript* variable. The python script assigned to this variable performs the following steps:

    - Loads the *pickle* package to handle interacting with the model

    - Uses the *pickle.loads()* function to deserialize the *bhmodel_serialized* model object and assign the results to *bhmodel_deserialized*

    - Uses the *predict()* method of the *bhmodel_deserialized* object to predict the medv using the *dfInputData* data frame

---

Note that you were able to include the whole data frame. That is because it only contains the columns needed by the regression model. In more advanced situations, you may need to perform transformations to the data passed to Python to get it in the structure that is needed by the model's formula.

    - Creates a *dfOutputData* data frame using the *dfInputData* data frame with *pred_medv* variable added as a new column

4. Sets the names you want to use for the input and output data frames using the @InputDFName and @OutputDFName variables.

5. Configures the *sp_execute_external_script* stored procedure. Most of parameters are self-explanatory, but explanations will be given for the *@params* and *@bhmodel_serialized* parameters. The *@params* argument enables you to define extra parameters needed by the model. In this case, you are defining a parameter named *@bhmodel_serialized* that has a *varbinary(max)* data type. This parameter will pass the model to the Python script and the model will be exposed to the script using the parameter name minus the @ sign. The value of the *@bhmodel_serialized* parameter is set in the succeeding line of code.

6. The *WITH RESULT SETS* clause is used to define the column names and column data types of the output. This information will be help Power BI determine the data types it should use and how to name the columns in the Power BI data model.

Execute the preceding script to add the stored procedure to the database. Now, you will be able to score the data by calling the stored procedure.

Step 6: Fetch the scored data in Power BI from SQL Server

To fetch the scored data in Power BI, launch Power BI, then go to *GetData* ➤ *SQL Server*. Populate the *Server and Database (optional)* textboxes with your server name and the name of the database that warehouses the model. Next, expand the *Advanced options* section and type EXEC [dbo].[uspPredictHousePrices_Python] in the *SQL statement (optional, requires database)* textbox. The *Include relationship columns* check box is checked by default, but it is not needed in this situation, so it is unchecked. Figure 10-4 shows what the form looks like after it has been configured.

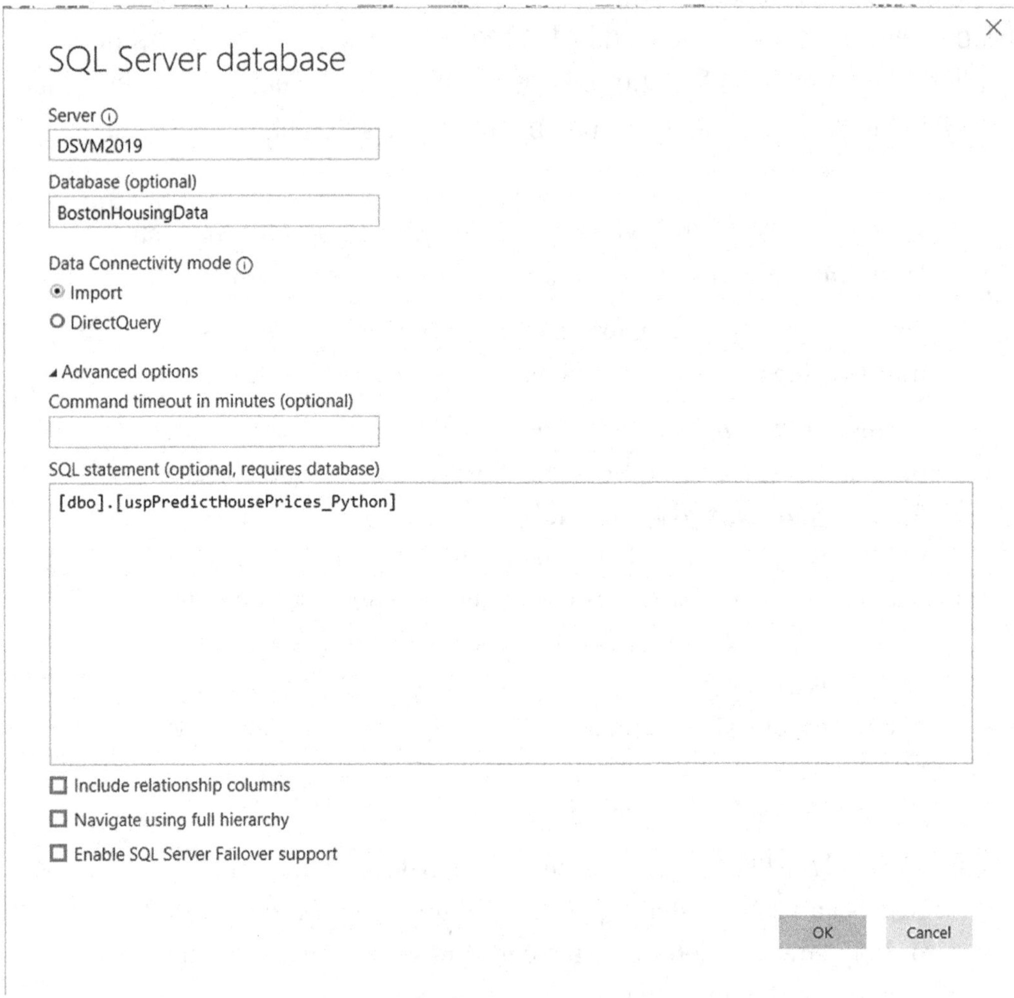

Figure 10-4. *GetData form populated with information needed to execute the dbo.uspPredictedHousePrices_Python stored procedure*

Click *OK* to add the data to the Power BI data model.

Just like in the R example, the focus in this example was not on the model but how to implement the model on Power BI data via *SSMLS*. Developing accurate machine learning models is well beyond the scope of this book. But if you or the data scientist on your team already knows how to develop complex models in Python that produce accurate predictions, then the steps outlined in this chapter can be used to apply those Python models to Power BI data models via *SSMLS*.

Performing sentiment analysis in Power BI using R in SQL Server Machine Learning Services

You used a Python script in Chapter 9 to call *Microsoft Cognitive Services* to perform sentiment analysis. Most DBAs don't allow API calls to be made from the database so *Microsoft Cognitive Services* will not be an option in most situations. Luckily, the pre-built models that come with *SSMLS* enable you to perform sentiment analysis without the need of making API calls to *Microsoft Cognitive Services*. Here are the steps needed to perform sentiment analysis using a pre-built *SSMLS* model.

Add pre-built R models to SQL Server Machine Learning Services using PowerShell

Step 1: Check to see if the pre-trained models are installed

Check to see if the pre-built models are installed in the following path:

```
C:\Program Files\Microsoft SQL Server\MSSQL15.MSSQLSERVER\R_SERVICES\
library\MicrosoftML\mxLibs\x64
```

You should see the following files in the preceding path:

- AlexNet_Updated.model

- ImageNet1K_mean.xml

- pretrained.model

- ResNet_101_Updated.model

- ResNet_18_Updated.model

- ResNet_50_Updated.model

If you do, you can skip to the *Use pre-built R sentiment model in SQL Server Machine Learning Services to score data in Power BI* section. Otherwise, go to Step 2.

Step 2: Open PowerShell as administrator

There are multiple ways to open PowerShell but an easy way is to search for PowerShell in the search bar next to the *Windows* icon on the taskbar. Doing so will cause the *Windows PowerShell* app to appear in the results. Right-click it and select *Run as administrator*.

Step 3: Download PowerShell script

Go to `https://aka.ms/mlm4sql` to download the file *Install-MLModels.ps1*. This file contains the PowerShell script needed to add the Python models to *SSMLS*. Clicking the link should cause the file to get downloaded to your *Downloads* folder. Verify that the file was successfully downloaded to the *Downloads* folder because the next step assumes that it will be there.

Step 4: Run the downloaded script in PowerShell

Run the following command in PowerShell:

```
C:\Users\<user-name>\Downloads\Install-MLModels.ps1 MSSQLSERVER
```

If the file was downloaded to your *Downloads* folder, then you will just needed to change the *<user-name>* to your username. Please refer to the "Troubleshooting" section if you are not able to successfully run the code.

Troubleshooting

- If you can't run the script, you may not have rights. You can see what rights you have by running the following code in PowerShell:

  ```
  Get-ExecutionPolicy
  ```

- If it is set to *restricted,* you can change it to *unrestricted* using the following code:

```
Set-ExecutionPolicy unrestricted
```

- After you run the preceding code, you should be able to run the *Install-MLModels.psi* script. PowerShell should be returned back to restricted state if that is what it was originally in. You can do so with the following code:

```
Set-ExecutionPolicy restricted
```

Use pre-built R sentiment model in SQL Server Machine Learning Services to score data in Power BI

If SQL Server Machine Learning Services is enabled and if you were successful at loading the pre-trained models, then you are in the position to be able to perform the sentiment analysis. Here are the steps you need to follow.

Step 1: Begin defining the stored procedures

Here is the T-SQL code that will be used to begin defining the stored procedure:

```
CREATE PROCEDURE [dbo].[getSentiments_R]
AS

BEGIN

END
```

The *CREATE PROCEDURE* command is used to add a stored procedure to a database. You are adding a stored procedure named *[dbo].[getSentiments_R]* in the preceding script. The [dbo] in the name represents the schema that the stored procedure belongs to. Schemas are used to group database objects together. The default schema is *dbo*. The second part is the actual name. After the stored procedure has been named, you type *AS* and follow that with the T-SQL code that defines the stored procedure. In this example, the T-SQL code will be wrapped in a *BEGIN... END* statement so that it will be executed in a batch.

Step 2: Define variables

Here are the variables that will be used in the stored proc:

```
DECLARE @RScript nvarchar(max);
DECLARE @Query nvarchar(max);
DECLARE @InputDataFrame nvarchar(128) = 'dfInput';
DECLARE @OutputDataFrame nvarchar(128) = 'dfOutput';
```

The *@RScript* variable will hold the R code that is responsible for performing the sentiment analysis, the *@Query* variable is used to hold the T-SQL statement that is used to define the input data set that will be passed to R, the *@InputDataFrame* variable will hold the name that R will use to refer to the input data set, and the *@OutputDataFrame* variable will hold the name that R will use to refer to the output data set.

Step 3: Set @Query variable

The following code is used to set the *@Query* variables:

```
SET @Query = 'SELECT [id], [text] ' +
             'FROM [dbo].[SentimentData]'
```

The code builds the T-SQL string needed to retrieve the input data for the script. The field that the sentiment analysis will be performed on is the *text* field. The *id* field is used to uniquely identify each record in the data set.

Step 4: Set @RScript variable

The @RScript variable is set below with the R code that will do the scoring:

```
SET @RScript = '
dfInput$text = as.character(dfInput$text)
sentimentScores <-
  rxFeaturize(
    data = dfInput,
    mlTransforms = getSentiment(
      vars = list(SentimentScore = "text"))
)

sentimentScores$text <- NULL
dfOutput <- cbind(dfInput, sentimentScores)'
```

354

The script is relatively short because the pre-trained model does the heavy lifting. In the preceding R script, the first line changes the *text* field in the *dfInput* data frame to a character data type. By default, R converts any character-based fields into a data type called a *factor*. Factor fields are used for categorical data.

Data stored as a factor data type is stored using a method similar to the dictionary enconding used in Microsoft's tabular engine. Each unique element in the field is replaced with an integer that serves as an index, and a map is created that maps the index to the unique element it replaces. This technique is beneficial because integers are more efficient to work with and have a lower memory footprint than long strings. Factors are good for data analysis, but the benefits you gain from using them are not needed in this example. You just need the actual text, so you convert the field to the character data type using the *as.character()* function from base R.

The actual sentiment analysis is performed in this section of the script:

```
sentimentScores <-
    rxFeaturize(
        data = dfInput,
        mlTransforms = getSentiment(
                    vars = list(SentimentScore = "text")
        )
    )
```

The workhorse functions are the *rxFeaturize()* and *getSentiment()* functions *from RevoScaleR. RevoScaleR* is a package of R functions built to overcome many of the challenges of working with big data sets.

RevoScaleR is a very powerful package, and I highly recommend you learn how to use it if you plan to do serious R development in *SSMLS*. Here is the URL to a guide that provides a thorough coverage of *RevoScaleR*: https:// packages.revolutionanalytics.com/doc/8.0.0/win/RevoScaleR_ Users_Guide.pdf.

The *rxFeaturize()* function is used to access data that has undergone a machine learning data transformation via *MicrosoftML*. It specifies the machine learning transformation that is being performed in the *mlTransforms* argument. The machine learning transformation that is occurring in this example is a sentiment transformation via *getSentiment()*. The *vars* argument in *getSentiment()* is used to specify the fields in the data set that the sentiment analysis will be performed on. If you specify a *named list,* then the *name* of each element in your list represents the name of the column that will hold the sentiment score, and the *value* represents the field you want to perform the sentiment analysis on. A nice benefit that you get from using a *named list* is that you can use a *named list* with multiple elements and score multiple fields at once. Here is what the code would look like if you were scoring two additional fields named *textB* and field *textC*:

```
sentimentScores <-
    rxFeaturize(
        data = dfInput,
        mlTransforms = getSentiment(
                    vars = list(
                            SentimentScore = "text",
                            SentimentScoreB = "textB",
                            SentimentScoreC = "textC")
        )
    )
```

The *getSentiment()* function will return a numeric value between 0 and 1 for each sentiment analysis it performs. The closer to 0 the value is, the more negative the sentiment, and the closer to 1 the value is, the more positive the sentiment. Typically, at this point you will come up with a rule to determine how you want to categorize the results. The most common method is to make the outcome binary by using a cutoff point that will be used to determine what is a good sentiment and what is a bad sentiment. The most common cutoff point is 0.5. When that cutoff point is used, any text that is scored 0.5 or greater is considered positive and anything that is scored less than 0.5 is considered negative.

You are not limited to converting your scores to a binary category. For instance, you may want to convert it into three categories. You may decide to say that any score between 0.33 and 0.66 is considered neutral, any score that is greater than 0.66 is good, and any score that is less than 0.33 is bad. The way you categorize the output is totally up to you.

Step 5: Configure sp_execute_external_script

Here is the code that is used to configure sp_execute_external_script:

```
EXEC sp_execute_external_script
      @language = N'R'
     ,@input_data_1 = @Query
     ,@input_data_1_name = @InputDFName
     ,@output_data_1_name = @OutputDFName
     ,@script = @RScript
```

The *sp_execute_external_script* special stored procedure is used to execute code in SQL Server that is using a programming language other than T-SQL. At the time of the writing, the available languages are R, Python, and Java. The parameters that are needed for this special stored procedure depend on the script you are executing. In this scenario, you need to define five parameters. Here are the parameters with descriptions:

- *@language* is used to specify the language you are using.

- *@input_data_1* represents the T-SQL code needed to create the input data set.

- *@input_data_1_name* represents the name that the R script will use to refer to the input data frame.

- *@output_data_1_name* represents the name that the R script will use to refer to the output data frame.

- *@script* will hold the R script that does the sentiment analysis.

Step 6: Define the output

The output is defined using the following code:

```
WITH RESULT SETS((
        [crim] float
       ,[rm] float
       ,[tax] float
       ,[lstat] float
       ,[pred_medv] float
));
```

Without the preceding code, SQL Server will return the output without names or known data types. This can be a problem when you are calling the stored procedure from Power BI. The *WITH RESULT SETS* clause enables you to give the output names and data types to make it more consumable by third-party clients like Power BI.

Step 7: Add the procedure to the database

Listing 10-6 contains the complete script.

Listing 10-6. The getSentiments_R stored procedure

```
CREATE PROCEDURE [dbo].[getSentiments_R]
AS

BEGIN

    DECLARE @RScript nvarchar(max);
    DECLARE @Query nvarchar(max);
    DECLARE @InputDFName nvarchar(128) = 'dfInput';
    DECLARE @OutputDFName nvarchar(128) = 'dfOutput';

    SET @Query = 'SELECT [id], [text], [likes] ' +
                 'FROM [dbo].[SentimentData]'
    SET @RScript = '
        dfInput$text = as.character(dfInput$text)
        sentimentScores <-
            rxFeaturize(
                data = dfInput,
                mlTransforms = getSentiment(
                vars = list(SentimentScore = "text"))
            )

        sentimentScores$text <- NULL
        dfOutput <- cbind(dfInput, sentimentScores)'
    '
```

```
EXEC sp_execute_external_script
    @language = N'R'
    ,@input_data_1 = @Query
    ,@input_data_1_name = @InputDFName
    ,@output_data_1_name = @OutputDFName
    ,@script = @RScript
  WITH RESULT SETS (
      (
       [id] [bigint],
       [text] [varchar](8000),
       [score] [float]
      )
   )

END
```

Run the preceding code in the *SentimentsDB* database in SQL Server to add the stored procedure to the database.

Step 8: Call the procedure from Power BI

To fetch the scored data in Power BI, launch Power BI, then go to *GetData* ➤ *SQL Server*. Populate the *Server and Database (optional)* textboxes. Next, expand the *Advanced options* section and type EXEC [dbo].[getSentiments_R] in the *SQL statement (optional, requires database)* textbox. The *Include relationship columns* check box is checked by default, but it is not needed in this situation, so it is unchecked. Figure 10-5 shows what the form looks like after it has been configured.

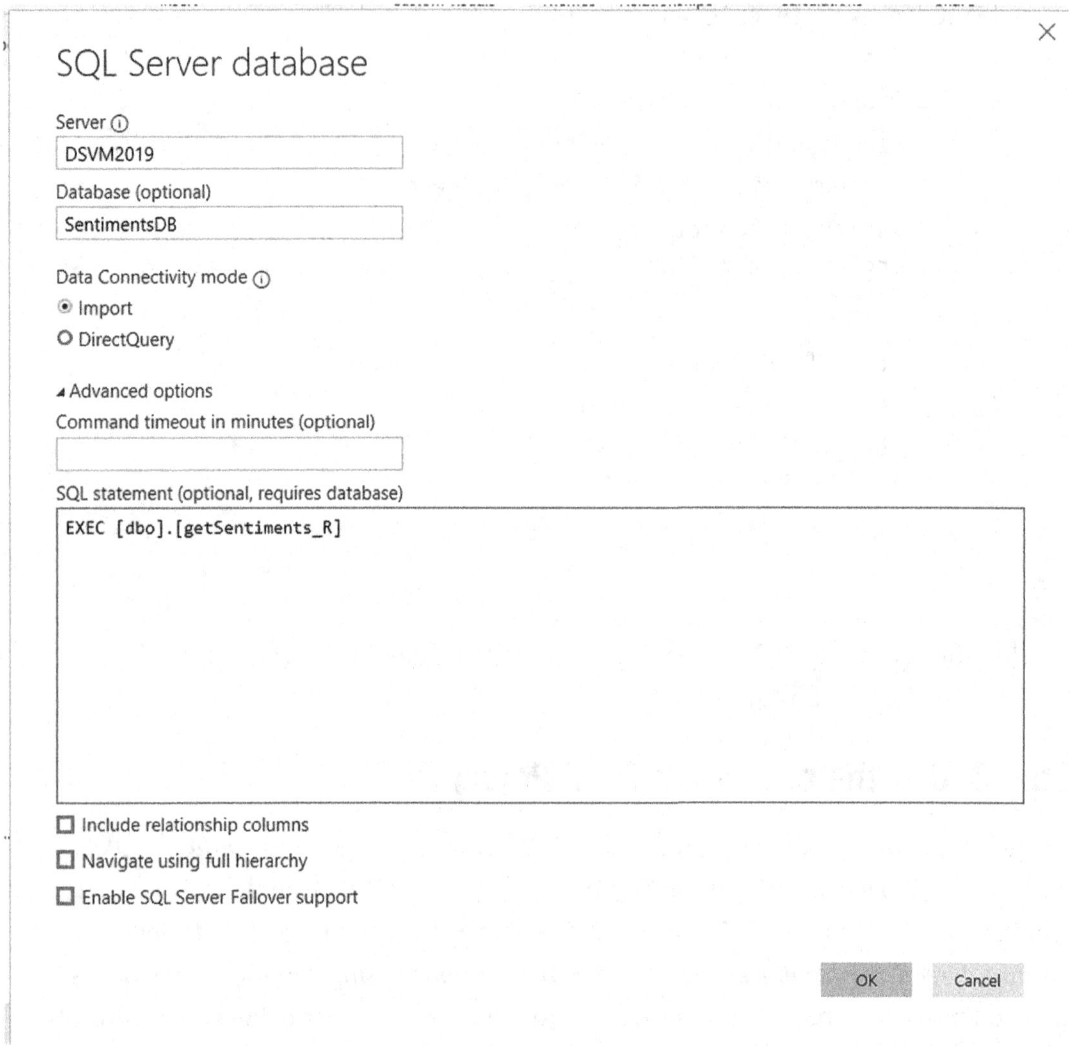

Figure 10-5. *GetData form populated with information needed to execute the dbo.getSentiment_R stored procedure*

As in similar stored procedure calls, the result will be exposed to the Power BI data model.

Performing sentiment analysis in Power BI using Python in SQL Server Machine Learning Services

As stated in the R version of this example, you can't use *Microsoft Cognitive Services* to perform sentiment analysis in *SSMLS* because of security reasons. But the sentiment pre-built model in SSMLS allows you to perform sentiment analysis without having to make API calls that may be a security risk. Here are the steps needed to perform sentiment analysis using the pre-built *SSMLS* model via Python.

Add pre-built Python models to SQL Server Machine Learning Services

Step 1: Check to see if the pre-trained models are installed

Check to see if the pre-built models are installed in the following path:

```
C:\Program Files\Microsoft SQL Server\MSSQL15.MSSQLSERVER\PYTHON_SERVICES\
Lib\site-packages\microsoftml\mxLibs
```

You should see the following files in the preceding path:

- AlexNet_Updated.model

- ImageNet1K_mean.xml

- pretrained.model

- ResNet_101_Updated.model

- ResNet_18_Updated.model

- ResNet_50_Updated.model

If you do, you can skip to the "Use pre-built Python sentiment model in SQL Server Machine Learning Services to score data in Power BI" section. Otherwise, go to Step 2.

Step 2: Open PowerShell as administrator

There are multiple ways to open PowerShell, but an easy way is to search for PowerShell in the search bar next to the *Windows* icon on the taskbar. Doing so will cause the *Windows PowerShell* app to appear in the results. Right-click it and select *Run as administrator*.

Step 3: Download PowerShell script

Go to `https://aka.ms/mlm4sql` to download the file *Install-MLModels.ps1*. This file contains the PowerShell script needed to add the models to *SSMLS*. Clicking the link should cause the file to get downloaded to your *Downloads* folder. Verify that the file was successfully downloaded to the *Downloads* folder because the next step assumes that it will be there.

Step 4: Run the downloaded script in PowerShell

Run the following command in PowerShell:

```
C:\Users\<user-name>\Downloads\Install-MLModels.ps1 MSSQLSERVER
```

If the file was downloaded to your *Downloads* folder, then you will need to change the `<user-name>` to your username. Please refer to the "Troubleshooting" section if you are not able to successfully run the code.

Troubleshooting

- If you can't run the script, you may not have rights. You can see what rights you have by running the following code in PowerShell:

  ```
  Get-ExecutionPolicy
  ```

- If it is set to *restricted,* you can change it to *unrestricted* using the following code:

  ```
  Set-ExecutionPolicy unrestricted
  ```

- After you run the preceding code, you should be able to run the *Install-MLModels.ps1* script. PowerShell should be returned back to restricted state if that is what it was originally in. You can do so with the following code:

```
Set-ExecutionPolicy restricted
```

The pre-trained models should now be added to *SSMLS*.

Use pre-built Python sentiment model in SQL Server Machine Learning Services to score data in Power BI

You should be in a good position to score the data if you successfully performed the previous steps. The steps needed to score the data are as follows.

Step 1: Begin defining the stored procedures

```
CREATE PROCEDURE [dbo].[getSentiments_Python]
AS

BEGIN

END
```

The *CREATE PROCEDURE* command is used to add a stored procedure to a database. You are adding a stored procedure named *[dbo].[getSentiments_Python]* in the preceding script. The *[dbo]* in the name represents the *schema* that the stored procedure belongs to. *Schemas* are used to group database objects. The default schema is *dbo*. The second part is the actual name. After the stored procedure has been named, you type *AS* and follow that with the T-SQL code that defines the stored procedure. The T-SQL code will be defined in a BEGIN... END statement so that it will be executed in a batch.

Step 2: Define variables

```
DECLARE @PythonScript nvarchar(max);
DECLARE @Query nvarchar(max);
DECLARE @InputDataFrame nvarchar(128) = 'dfInput';
DECLARE @OutputDataFrame nvarchar(128) = 'dfOutput';
```

The *@PythonScript* variable will hold the Python code that is responsible for performing the sentiment analysis, the *@Query* variable is used to hold the T-SQL statement that is used to define the input data set that will be passed to Python, the *@InputDataFrame* variable will hold the name that Python will use to refer to the input data set, and the *@OutputDataFrame* variable will hold the name that Python will use to refer to the output data set.

Step 3: Set @Query

The @Query variable holds the T-SQL string that is needed to retrieve the input data for the Python script. Here is the code:

```
SET @Query = 'SELECT [id], [text] FROM [dbo].[SentimentData]'
```

The *@Query* variable holds the T-SQL string that is needed to retrieve the input data for the Python script.

Step 4: Set @PythonScript

The python code that will perform the sentiment analysis is set to the @PythonScript using the code below:

```
    SET @PythonScript = '
from microsoftml import rx_featurize, get_sentiment

sentiment_scores = rx_featurize(
    data=dfInput,
    ml_transforms=[get_sentiment(cols=dict(scores="text"))])

dfOutput = sentiment_scores'
```

The first line of code in the preceding script loads the required functions. They are the *rx_featurize()* function and *get_sentiment()* function from the *microsoftml* Python library. The next line uses the *rx_featurize()* function and *getSentiment()* function together to do the sentiment analysis. The *rxFeaturize()* function enables you to access data that has undergone a machine learning data transformation via *microsoftml*. It specifies the machine learning transformation that is being performed in the *mlTransforms* argument. In this example, the transformation is a sentiment transformation done using the *getSentiment()* function. The *cols* argument in *getSentiment()* is used to specify the columns in the data set that you want to score. It uses a dictionary, with the *key* of each *key/value* pair in the dictionary representing the name of the new column that will contain the sentiment scores and the *value* representing the column that contains the text you are scoring.

Please note that you can perform sentiment analysis on multiple columns at once. You just need to add the necessary key/value pairs to your dictionary with the new keys representing the new columns that will hold the sentiment scores and the new values representing the columns you want to score.

The *getSentiment()* function will return a numeric value between 0 and 1 for each sentiment analysis it performs. The closer to 0 the value is, the more negative the sentiment, and the closer to 1 the value is, the more positive the sentiment.

Step 5: Configure sp_execute_external_script

The *sp_execute_external_script* is configured using the following code:

```
EXEC sp_execute_external_script
        @language = N'Python'
        ,@input_data_1 = @Query
        ,@input_data_1_name = @InputDFName
        ,@output_data_1_name = @OutputDFName
        ,@script = @PythonScript
```

In this step, you configure the *sp_execute_eternal_script* procedure. The configuration is almost identical to the one that we did for R. The only difference is the language parameter.

Step 6: Define the output

The output is defined using the following code:

```
WITH RESULT SETS((
        [crim] float
        ,[rm] float
        ,[tax] float
        ,[lstat] float
        ,[pred_medv] float
));
```

Without the preceding code, SQL Server will return the output without names or known data types. This can be a problem when you are calling the stored procedure from Power BI. The *WITH RESULT SETS* clause enables you to give the output columns names and data types to make it more consumable by third party clients like Power BI.

Step 7: Add the procedure to the database

Listing 10-7 contains the complete script.

Listing 10-7. The getSentiments_Python stored procedure

```
CREATE PROCEDURE [dbo].[getSentiments_Python]
AS

BEGIN

        DECLARE @PythonScript nvarchar(max);
        DECLARE @Query nvarchar(max);
        DECLARE @InputDFName nvarchar(128) = 'dfInput';
        DECLARE @OutputDFName nvarchar(128) = 'dfOutput';

        SET @Query = 'SELECT ID, [text] FROM dbo.SentimentData'

        SET @PythonScript = '
import pandas as pd
from microsoftml import rx_featurize, get_sentiment

sentiment_scores = rx_featurize(
    data=dfInput,
    ml_transforms=[get_sentiment(cols=dict(scores="text"))])

dfOutput = sentiment_scores
'
```

```
EXEC sp_execute_external_script
      @language = N'Python'
      ,@input_data_1 = @Query
      ,@input_data_1_name = @InputDFName
      ,@output_data_1_name = @OutputDFName
      ,@script = @PythonScript

WITH RESULT SETS (
      (
      [ID] [bigint],
      [text] [varchar](8000),
      [score] [float]
      )
   )

END
```

Run the preceding T-SQL code via *SSMS* in the *SentimentsDB* to add the *[dbo].*
[getSentiments_Python] to the *SentimentsDB* database.

Step 8: Call the procedure from Power BI

To score the data, launch Power BI, then go to *GetData* ➤ *SQL Server*. Populate the
Server and Database (optional) textboxes. Next, expand the *Advanced options* section
and type EXEC [dbo].[getSentiments_Python] in the *SQL statement (optional, requires
database)* textbox. The *Include relationship columns* check box is checked by default,
but it is not needed in this situation, so it is unchecked. Figure 10-6 shows what the form
looks like after it has been configured.

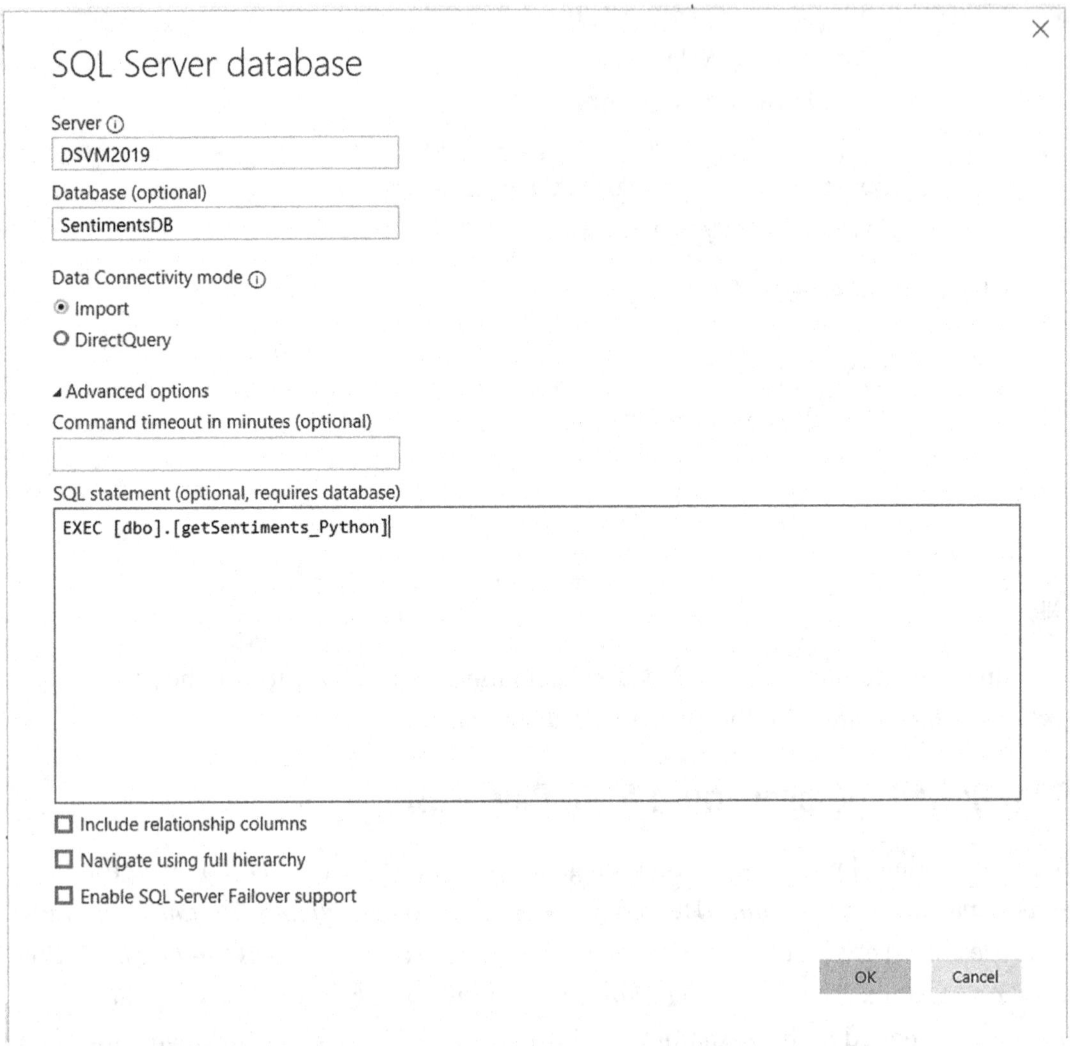

Figure 10-6. *GetData form populated with information needed to execute the dbo.getSentiment_Python stored procedure*

The results will be available to the Power BI data model.

Calculating distance in Power BI using R in SQL Server Machine Learning Services

There are more benefits from having access to R in SQL Server than scoring data using machine learning models. Some of those benefits include performing data transformations or performing advanced mathematical calculations that are hard to do with traditional T-SQL. Here is an example of performing an advanced mathematical calculation that is hard to implement in T-SQL but easy to implement in R. You will learn how to implement the *Haversine* formula to calculate distance between two geographical points. Let's go over the steps!

Step 1: Make sure dplyr is loaded in SSMLS

The *dplyr* package will be available if you are using the *DSVM*. If you are not, refer to the book's introduction for information on how to add R packages to *SSMLS*. You can run the following script shown in Listing 10-8 to get the available R packages in *SSMLS*.

Listing 10-8. Script to get installed R packages

```
EXECUTE sp_execute_external_script
  @language=N'R',
  @script = N'
packagematrix <- installed.packages();
Name <- packagematrix[,1];
Version <- packagematrix[,3];
OutputDataSet <- data.frame(Name, Version);'

WITH RESULT SETS ((PackageName nvarchar(250), PackageVersion nvarchar(max) ))
```

Step 2: Launch SSMS and connect to a SQL Server

Make sure to connect to a *SQL Server* that has *SSMLS* with the R enabled. It will be enabled by default if you are using the *DSVM*. If you are using an instance configured outside of *DSVM,* then it will need to be enabled. Please refer to the installation section of the book's introduction to get information about how to enable R in *SSMLS* if the *SQL Server* instance that you are working on does not have R enabled. You can run the following code as a quick test to see if it is:

369

```
EXEC sp_execute_external_script @language = N'R',
@script = N'
Test <- "Test R"
OutputDataSet = data.frame(Test)'
```

If R is enabled, a single cell table with the value *Test R* will be returned.

Step 3: Add the CalculateDistance database to the server

The *CalculateDistance* database is pre-configured with the data set and tables needed for this example. It is included in the book's repo. You will find it in the *Databases* folder for Chapter 10. Here are the steps you need to take in order to add the database to SQL Server:

1. Copy the *CalculateDistance.bak* file to the *Backup* folder for SQL Server. The location in the *DSVM* is C:\Program Files\Microsoft SQL Server\MSSQL15.MSSQLSERVER\MSSQL\Backup.

2. Go to SSMS and right-click the *Databases* folder, then select *Restore Databases....*

3. The *Restore Database* pop-up form will appear. Make sure you are on the *General* page tab. Select the *Device* radio button in the *Source* section.

4. Click the eclipse button, the button with the three dots, then click the *Add* button, then browse to the location where *CalculateDistance.bak* is located.

5. Select the *CalculateDistance.bak* file, then click the *OK* button to close out the *Locate Backup File* form.

6. Click *OK* to close out the *Select backup devices* form.

7. Click *OK* to close out the *Restore Database* form. Doing so will add the database.

Step 4: Add the stored procedure that will calculate the distances

The T-SQL needed to create the stored procedure that will calculate the distance between the two addresses is shown in Listing 10-9.

Listing 10-9. The calcDistance_R stored procedure

```
CREATE PROCEDURE [dbo].[calcDistance_R]
AS

BEGIN

        DECLARE @RScript nvarchar(max);
        DECLARE @Query nvarchar(max);
        DECLARE @InputDFName nvarchar(128) = 'dfInput';
        DECLARE @OutputDFName nvarchar(128) = 'dfOutput';

        SET @Query = 'SELECT [Employee_ID], [EmployeeAddress],
        [TerminalAddress], [lon_EmployeeAddress], [lat_EmployeeAddress],
        [lon_TerminalAddress], [lat_TerminalAddress]
                    FROM [dbo].[EmployeeList]'

        SET @RScript = '
                library(dplyr)

                ComputeDist <-
                        function(addressA_long, addressA_lat, addressB_long,
                        addressB_lat) {
                                R <- 6371 / 1.609344 #radius in mile
                                delta_lat <- addressB_lat - addressA_lat
                                delta_long <- addressB_long - addressA_long
                                degrees_to_radians = pi / 180.0
                                a1 <- sin(delta_lat / 2 * degrees_to_radians)
                                a2 <- as.numeric(a1) ^ 2
                                a3 <- cos(addressA_lat * degrees_to_radians)
                                a4 <- cos(addressB_lat * degrees_to_radians)
                                a5 <- sin(delta_long / 2 * degrees_to_radians)
                                a6 <- as.numeric(a5) ^ 2
```

```r
                        a <- a2 + a3 * a4 * a6
                        c <- 2 * atan2(sqrt(a), sqrt(1 - a))
                        d <- R * c
                        return(d)
                }

        dfOutput <-
        dfInput %>%
        mutate(
                Distance =
                        round(ComputeDist(lon_EmployeeAddress,
                        lat_EmployeeAddress, lon_TerminalAddress,
                        lat_TerminalAddress), 1)
        )'

EXEC sp_execute_external_script
        @language = N'R'
        ,@input_data_1 = @Query
        ,@input_data_1_name = @InputDFName
        ,@output_data_1_name = @OutputDFName
        ,@script = @RScript

WITH RESULT SETS (
        (
        [ID] [bigint],
        [AddressA] [varchar](50),
        [AddressB] [varchar](50),
        [lon_AddressA] [decimal](10,8),
        [lat_AddressA] [decimal](10,8),
        [lon_AddressB] [decimal](10,8),
        [lat_AddressB] [decimal](10,8),
                [Distance] [decimal](4,1)
        )
    )

END
```

The script uses the same logic that was used in Chapter 8. The method in Chapter 8 is good for self-service situations but not enterprise solutions. You are limited to using the personal version of the *on-premises data gateway* for refreshes when you use R scripts in Power BI. That limitation prevents you from using them in an enterprise solution. The situation changes when you use R in *SSMS*. You are able to use the enterprise version of the *on-premises data gateway* because your R script is embedded in a T-SQL stored procedure.

Now that you have the background about some of the benefits of using R-based stored procedures in *SSMLS*, let's go over how the preceding stored procedure works:

1. Defines the following variables needed in the script:

    - *@RScript* variable holds the R script that does the calculation.

    - *@Query* variable holds the TSQL script that defines the data set that the calculation will be applied on.

One of the benefits of having the input data set defined by a T-SQL statement is that you can use the power of T-SQL to incorporate complex business logic or to perform complex data transformations to define the data set.

    - *@InputDFName* variable holds the name of the input data set that R will use to refer to the input data set.

    - *@OutputDFName* variable holds the name of the output data set that R will use to refer to the output data set.

2. Sets the *@Query* variable with the TSQL code that returns the input data set.

3. Sets the *@RScript* variable. The R script assigned to this variable performs the following steps:

    - Loads the *dplyr* package. This package is used to add the column to the data set with the computed distance.

    - Defines the *ComputeDist()* function. This is the same function that we used in Chapter 8. We are just using the R script in a stored procedure instead of calling it from Power Query.

- Creates the *dfOutput* data frame that is based on the *dfInput* data frame that was passed to the R script, then adds a new column named *Distance* that is based on the *ComputeDist* function.

- Configures *sp_execute_external_script* using the variables we defined earlier.

- Configures the *WITH RESULT SETS* to explicitly define the column names and data types of the output data.

The preceding steps were explained in depth in Chapter 8. Please refer to that chapter if you need a more detailed explanation. Execute the T-SQL script in *SSMS* to add the *dbo.calcDistance_R* to the database.

Step 5: Call the Power BI procedure from Power BI

To score the data via Power BI, launch Power BI, then go to *GetData* ➤ *SQL Server.* Populate the *Server and Database (optional)* textboxes. Next, expand the *Advanced options* section and type EXEC [dbo].[calcDistance_R] in the *SQL statement (optional, requires database)* textbox. The *Include relationship columns* check box is checked by default, but it is not needed in this situation, so it is unchecked. Figure 10-7 shows what the form looks like after it has been configured.

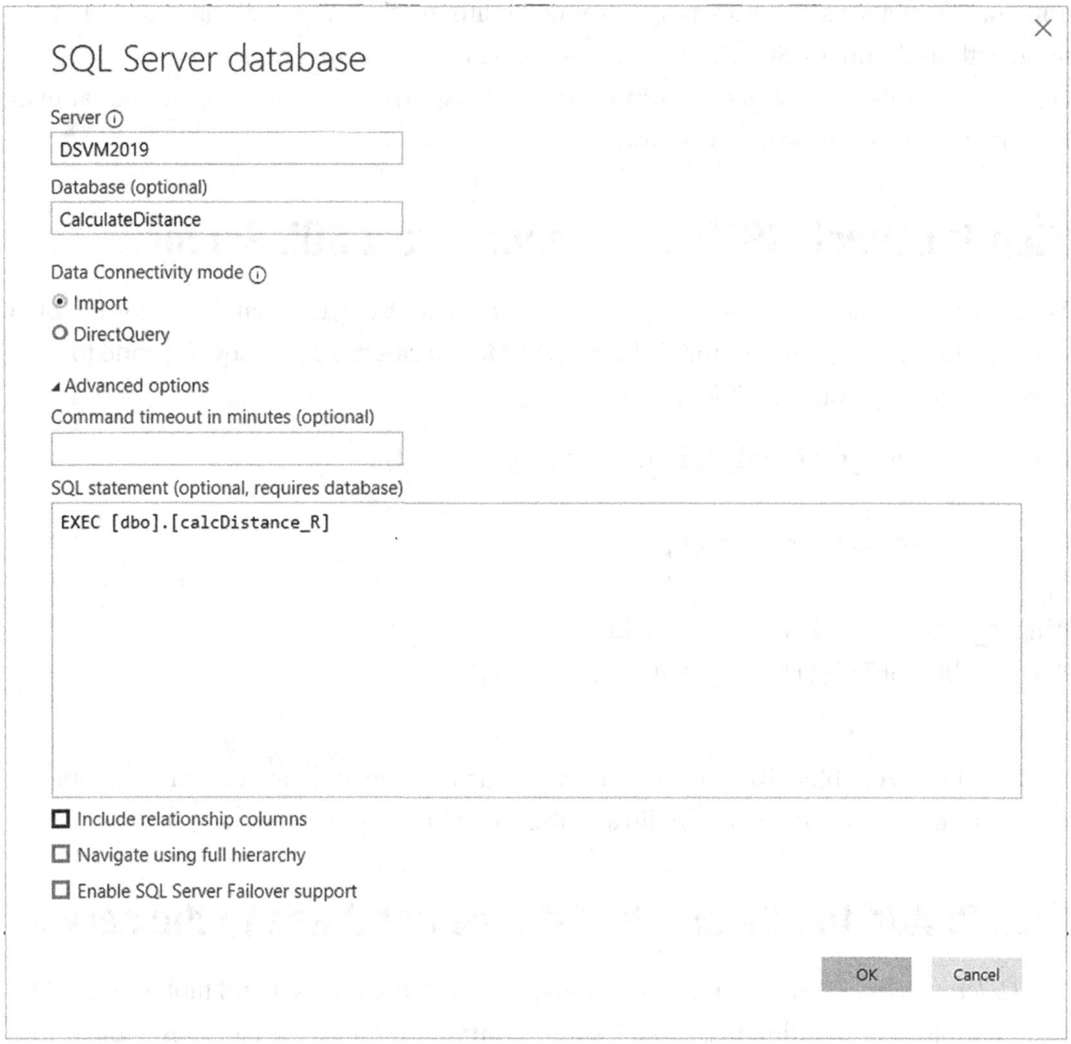

SQL Server database   ×

Server ⓘ

```
DSVM2019
```

Database (optional)

```
CalculateDistance
```

Data Connectivity mode ⓘ

◉ Import

○ DirectQuery

◢ Advanced options

Command timeout in minutes (optional)

```

```

SQL statement (optional, requires database)

```
EXEC [dbo].[calcDistance_R]



```

☐ Include relationship columns

☐ Navigate using full hierarchy

☐ Enable SQL Server Failover support

                                        OK   Cancel

Figure 10-7.   *GetData form populated with information needed to execute the dbo.calcDistance_R stored procedure*

The resulting data set will be available to the Power BI data model.

Calculating distance in Power BI using Python in SQL Server Machine Learning Services

Just like with R, there are more benefits from having access to Python in SQL Server than scoring data using machine learning models. Some of those benefits include performing data transformations or performing advanced mathematical calculations that are hard to do with traditional T-SQL. You just completed an exercise that refactored a relatively complex formula, the *Haversine* formula, using R in *SSMLS*. Now you will do the same in Python. Let's go over the steps of how to do so.

Step 1: Launch SSMS and connect to a SQL Server

Make sure your instance of SQL Server has *SSMLS* with the Python enabled. It is enabled in the instance of SQL Server that is in the *DSVM*. You can use the following code to check to see if Python is enabled:

```
EXEC sp_execute_external_script @language =N'Python',
@script=N'
OutputDataSet = InputDataSet;
',
@input_data_1 =N'SELECT 1 AS hello'
WITH RESULT SETS ((([hello] int not null));
GO
```

If Python is enabled, the above script will return a table with one column and one row. The column header will be hello and the value in the row will be 1.

Step 2: Add the CalculateDistance database to the server

The *CalculateDistance* database is pre-configured with the data set and tables needed for this example. It is included in the book's repo. You will find it in the *Databases* folder for Chapter 10. Here are the steps you need to take to add the database to SQL Server:

1. Copy the *CalculateDistance.bak* file to the *Backup* folder for SQL Server. The location in the *DSVM* is C:\Program Files\Microsoft SQL Server\MSSQL15.MSSQLSERVER\MSSQL\Backup.

2. Go to *SSMS* and right-click the *Databases* folder, then select *Restore Databases....*

3. The *Restore Database* pop-up form will appear. Make sure you are on the *General* page tab. In the *Source* section, select the *Device* radio button.

4. Click the eclipse button, the button with the three dots, then click the *Add* button, then browse to the location where *CalculateDistance.bak* is located.

5. Select the *CalculateDistance.bak* file, then click the *OK* button to close out the *Locate Backup File* form.

6. Click *OK* to close out the *Select backup devices* form.

7. Click *OK* to close out the *Restore Database* form. Doing so will add the database.

Step 3: Add the stored procedure that will calculate the distances

The T-SQL needed to create the stored procedure that will calculate the distance between the two addresses is listed in Listing 10-10.

Listing 10-10. The calcDistance_Python stored procedure

```
CREATE PROCEDURE [dbo].[calcDistance_Python]
AS

BEGIN

    DECLARE @PythonScript nvarchar(max);
    DECLARE @Query nvarchar(max);
    DECLARE @InputDFName nvarchar(128) = 'dfInput';
    DECLARE @OutputDFName nvarchar(128) = 'dfOutput';

    SET @Query = 'SELECT [Employee_ID], [EmployeeAddress],
    [TerminalAddress], [lon_EmployeeAddress], [lat_EmployeeAddress],
    [lon_TerminalAddress], [lat_TerminalAddress]
                FROM [dbo].[EmployeeList]'
```

```
    SET @PythonScript = '
from math import cos, sin, atan2, pi, sqrt, pow

def ComputeDist(row):
    R = 6371 / 1.609344 #radius in mile
    delta_lat = row["lat_EmployeeAddress"] - row["lat_TerminalAddress"]
    delta_lon = row["lon_EmployeeAddress"] - row["lon_TerminalAddress"]
    degrees_to_radians = pi / 180.0
    a1 = sin(delta_lat / 2 * degrees_to_radians)
    a2 = pow(a1,2)
    a3 = cos(row["lat_TerminalAddress"] * degrees_to_radians)
    a4 = cos(row["lat_EmployeeAddress"] * degrees_to_radians)
    a5 = sin(delta_lon / 2 * degrees_to_radians)
    a6 = pow(a5,2)
    a = a2 + a3 * a4 * a6
    c = 2 * atan2(sqrt(a), sqrt(1 - a))
    d = R * c

    return d

dfOutput = dfInput

dfOutput["Distance"] = dfOutput.apply(lambda row: ComputeDist(row), axis=1)'
    EXEC sp_execute_external_script
            @language = N'Python'
            ,@input_data_1 = @Query
            ,@input_data_1_name = @InputDFName
            ,@output_data_1_name = @OutputDFName
            ,@script = @PythonScript

    WITH RESULT SETS (
            (
            [ID] [bigint],
            [AddressA] [varchar](50),
            [AddressB] [varchar](50),
            [lon_AddressA] [FLOAT],
```

```
    [lat_AddressA] [FLOAT],
    [lon_AddressB] [FLOAT],
    [lat_AddressB] [FLOAT],
        [Distance] [FLOAT]
    )
  )
```

END

The script uses the same logic that was used in Chapter 8. The method in Chapter 8 is good for self-service scenarios that is being maintained by an individual analyst. This method is better for enterprise solutions for the same reasons given in the R version of the exercise.

Now, let's go over how the preceding stored procedure works:

1. Defines the following variables needed in the script:

   - *@PythonScript* variable holds the Python script that performs the calculation.

   - *@Query* variable holds the TSQL script that defines the data set that the calculation will be applied on.

   - *@InputDFName* variable holds the name of the input data set that Python will use to refer to the input data set.

   - *@OutputDFName* variable holds the name of the output data set that Python will use to refer to the output data set.

2. Sets the *@Query* variable with the TSQL code that returns the input data set.

3. Sets the *@PythonScript* variable. The python script assigned to this variable performs the following steps:

   - Loads the *cos, sin, atan2, pi, sqrt,* and *pow* functions from the *math* library. Those functions will be used by the distance calculation formula in the script.

   - Defines the *ComputeDist()* function. This is the same function that we used in Chapter 8. We are just using it in a stored procedure instead of Power Query.

- Creates the *dfOutput* data frame based on the *dfInput* data frame that was passed to the script.

- Add the *Distance* field to the *dfOutput* data frame by using a *lambda* function to apply the *ComputeDist()* function.

4. Configures *sp_execute_external_script*. Note that the parameters in this task are different than the *Boston Housing* example. We are not using a model to score data in this task, so the parameters needed to handle models were not required.

5. Configures the *WITH RESULT SETS* clause so that you can explicitly define the column names and data types of your output data.

The preceding steps were explained in depth in Chapter 8. Please refer to that chapter if you need a more detailed explanation. Execute the T-SQL script in *SSMS* to add the *dbo.calcDistance_Python* to the database.

Step 4: Call the Power BI procedure from Power BI

To score the data in Power BI, launch Power BI, then go to *GetData* ➤ *SQL Server*. Populate the *Server and Database (optional)* textboxes. Next, expand the *Advanced options* section and type [dbo].[calcDistance_Python] in the *SQL statement (optional, requires database)* textbox. The *Include relationship columns* check box is checked by default, but it is not needed in this situation, so it is unchecked. Figure 10-8 shows what the form looks like after it has been configured.

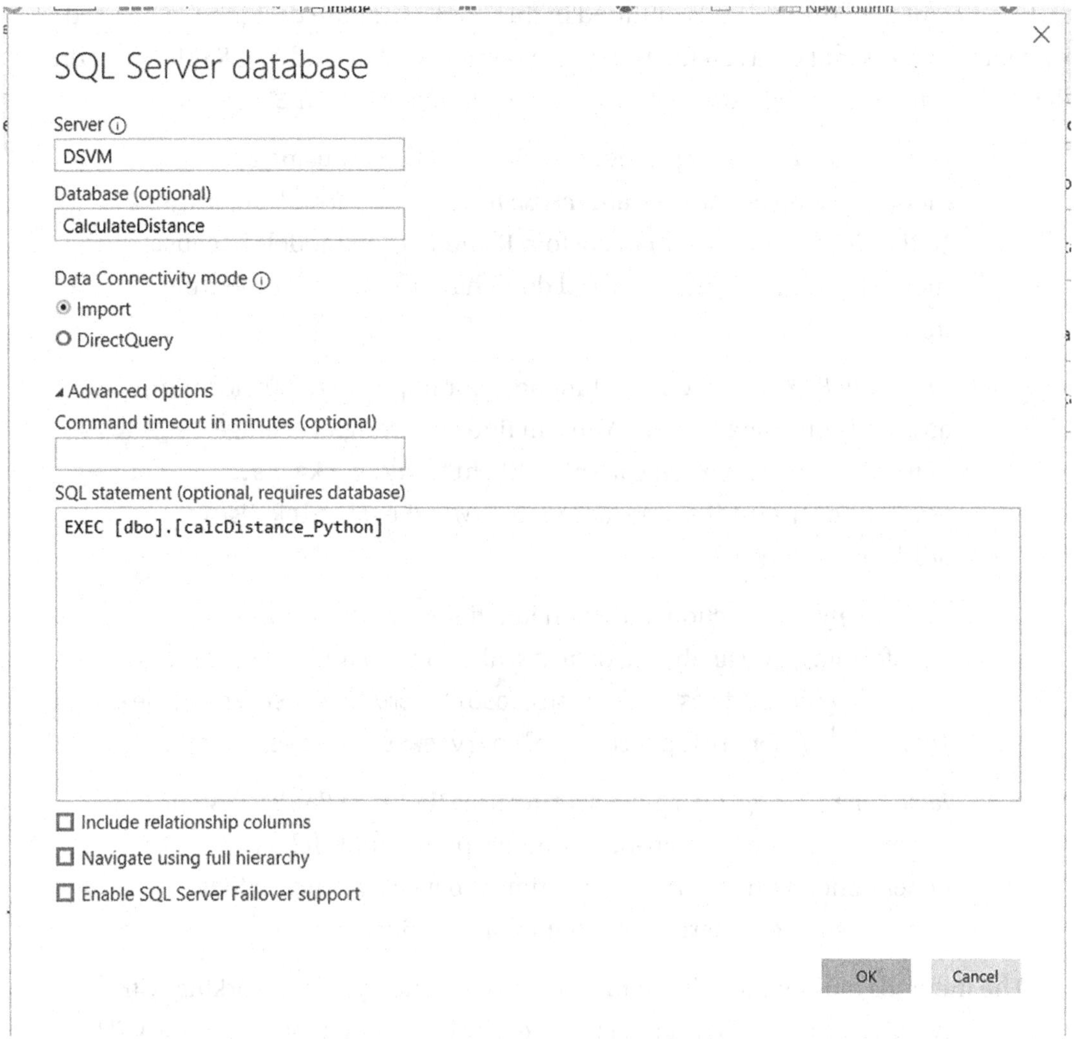

SQL Server database

Server ⓘ

```
DSVM
```

Database (optional)

```
CalculateDistance
```

Data Connectivity mode ⓘ

◉ Import

○ DirectQuery

◢ Advanced options

Command timeout in minutes (optional)

SQL statement (optional, requires database)

```
EXEC [dbo].[calcDistance_Python]
```

☐ Include relationship columns

☐ Navigate using full hierarchy

☐ Enable SQL Server Failover support

OK   Cancel

Figure 10-8. *GetData form populated with information needed to execute the dbo.calcDistance_Python stored procedure*

In this chapter, you learned how to productionize your R and Python scripts for use with Power BI via *SSMLS*. They remained accessible to Power BI because they were productionized by wrapping them in a special stored procedure in SQL Server. The coverage given to *SSMLS* was very high level. A thorough coverage of *SSMLS* would require its own book. *SSMLS* has tools that make doing data science on big data sets more efficient than the methods outlined in this book. They are designed to be used by data engineers and data scientists. If you are interested in studying *SSMLS* in more depth, I highly recommend that you research the following *SSMLS* topics:

- *Native Scoring*: Makes predictions via a T-SQL view using a trained model developed with certain *revoscalepy* or *RevoScaleR* algorithms. Native Scoring is faster than custom R and Python models because they use native C++ libraries and don't have to call an R or Python interpreter.

- *RevoScaleR*: A collection of R functions for importing, transforming, and analyzing data at scale. You can find a thorough coverage of the RevoScaleR package at this URL: `https://packages.revolutionanalytics.com/doc/8.0.0/win/RevoScaleR_Users_Guide.pdf`.

- *revoscalepy*: A collection of Python functions for importing, transforming, and analyzing data at scale. You can find out more about *revoscalepy* here: `https://docs.microsoft.com/en-us/sql/machine-learning/python/ref-py-revoscalepy?view=sql-server-ver15`.

- *Real Time Scoring*: Real-time scoring uses the *sp_rxPredict* system stored procedure for high-performance predictions. It has no dependencies on R or Python runtimes so it can be run on SQL Server instances where R and Python are not installed.

The methods outlined in this chapter work great when you are working with relatively small SQL Server data sets that can easily fit in your server's memory. The above features have functionality that you can leverage when your data is too big to fit into memory.

In addition to *SSMLS*, Microsoft offers cloud-based enterprise solutions that you can leverage as well from Power BI. These solutions include *Azure Machine Learning Services*, *Azure Databricks*, and *Azure Synapse,* to name a few. Even more solutions are available if you are using Power BI Premium. This book purposely focused on the *SSMLS* because it is freely available to those who have SQL Server 2016 and later and does not require you to have premium compacity. Maybe the cloud-based solutions will be covered in the next version of this book!

I really appreciate you purchasing the book and I hope you find the book's content very useful and educational!

Index

A

AdventureWorksDW_StarSchema
database, 191, 192, 195, 209
aes() function, 1, 3, 10, 20, 22, 26, 32–34,
72, 104, 116, 117, 131, 144
all.equal() function, 46, 69, 70, 84,
96, 128, 143
annotate() function, 56, 57, 133, 134
Artificial intelligence (AI), 295
as_data_frame() function, 202

B

bhmodel_deserialized variable, 337
@bhmodel_serialized
parameters, 338, 349
Bubble chart, 39
add lables, 74, 75
chart titles, 71
colors, 76, 77
conferences/divisions colors, 70, 71
data set, 67, 71, 72
data validation test, 70
dimensions of data, 66, 67, 72
filter slicer, 68
geom_point() geom, 73, 74
ggplot() function, 72
ggtitle() function, 77, 78
load data, 68
packages, 69

R script, 78–80
R visual configuration, 68
variables, 69

C

Callout chart, 39
add columns, 47
align titles, 62, 63
axis, 54, 55
column chart layer, R visual, 52
data set, 41, 42
data validation test, 47
dynamic annotation, 56–58
dynamic titles/caption, 58, 59
export data, R Studio, 43–45
horizontal bar chart/dynamic
annotations, 40
packages, 45
remove labels, 59, 60
remove legend, 60, 61
R script, 64–66
R visual configuration, 43
slicer, 42, 43
text layer, R visual, 52–54
theme_few() theme, 61, 62
variables, dynamic
portions, 46, 48, 50, 51
chdir() function, 164, 249, 255
colnames() function, 46

G, H

I

J

K

L

Printed in the United States
By Bookmasters